MIT GEOPHYSICAL ANALYSIS GROUP

Volume 5 in the "Scientist and Science" series

Enders Anthony Robinson

Professor Emeritus in the
Maurice Ewing and J. Lamar Worzel Chair
Columbia University in the City of New York

Goose Pond Press

Available from Amazon.com and other retail outlets.

This book is dedicated to
CHARLES CROSS
of Alford in Lincolnshire and San Francisco

Goose Pond Press

Ralph Waldo Emerson told Henry David Thoreau that Goose Pond should be called The Droplet or God's Pond. It was significant to many of Concord's leading literary figures, all of whom walked there often.

Contents

Preface

In 1953 the MIT Department of Geology and Geophysics and the MIT Mathematics Department came to the rescue of the academic use of the MIT Whirlwind digital computer. It was essential for the geophysical application of Prof. Norbert Wiener's mathematics. Prof. Robert Shrock (*Geology at MIT 1865-1965*, Volume 2, MIT Press, 1982) writes:

> Sometime in 1953, as I was becoming acquainted with the duties of a Department Head, I was asked to meet with J. A. Stratton [the Vice President of MIT], along with all other Heads of both Departments and Laboratories, to consider the question---"Will MIT have enough future demand for a high-speed digital computer, specifically WHIRLWIND, to justify keeping it, or should it be disposed of?" At the time it was not surprising that there was a question about the future of high-speed computers; even experienced scientists and engineers did not foresee what was to come. It was a surprise to the group, however, when it came my turn and I stated that our Department of Geology and Geophysics could use a large block of time. For what could we in geology possibly use the many hours that Professor George Wadsworth [of the Mathematics Department] had advised me to report? The answer, of course, was the **Geophysical Analysis Group (GAG)** Project that was well underway. I was asked to describe the proposed computation needs and sent the request to Stratton. What happened at that meeting led to MIT's decision to keep the WHIRLWIND computer for the time being, but meanwhile to plan for better and faster machines. Soon came the electronic digital computers--the IBM 704s, 709s, and 7090s, and the rest of the rapid development of computation at MIT is now history. The GAG Project could not have been carried out without WHIRLWIND. But it was not at all certain in the early 1950s that MIT would need the kinds of computers that would become necessities in the next two decades.

This book goes back seventy years to 1946 when Prof. George Valley taught a section of first year physics at MIT. If anyone deserves credit as the father of the Internet, it is he. On visits to Whirlwind in 1950 Prof. Valley realized that the solution to air defense would reside in the digital computer. Valley was instrumental in the formation of the MIT Lincoln Laboratory. The result was the Semi-Automatic Ground Environment (SAGE) Air Defense System. The massive scope of SAGE came at enormous cost. Its technology and many of its 10,000 computer programmers were later transferred to ARPA. The result was ARPANET. The Internet sprang from ARPANET.

The MIT Geophysical Analysis Group (GAG) was initiated by three dedicated professors: Prof. George Wadsworth and Prof. Norbert Wiener of the Mathematics Department and Prof. Patrick Hurley of the Geology and Geophysics Department. In 1950 I was a research assistant in mathematics. They assigned me the task of directing the newly-formed geophysics project. The problem was to find a better way to find oil. The solution would reside in the digital computer. I was designated Director of the Geophysical Analysis Group in 1952 when it was officially established within the MIT Department of Geology and Geophysics. The Department has now become the much more comprehensive MIT Department of Earth, Atmospheric and Planetary Sciences (EAPS). It deals with the inaccessible inner Earth, turbulent oceans, unsettled atmosphere, distant planets, and origins of life. MIT's research projects and educational initiatives bring new understanding of the workings of the world. The need for financial support for EAPS is as great as ever.

I want to thank Schlumberger Prof. Robert van der Hilst, Head of MIT Department of Earth, Atmospheric and Planetary Sciences; Lee Lawyer, SEG, Tulsa; Prof. M. Nafi Toksöz, EAPS; Norbert Wiener Prof. David Vogan, MIT Mathematics Department; Monique Gardiner, Carmel, CA; Ralph Warrington, Shell Oil retired; Distinguished Prof. Donald Gaver, Naval Postgraduate School, Monterey, CA; Prof. Norman Ness, University of Delaware and NASA space scientist; Dr. David M. Rubin, Optometry, Venice, Florida; Angela Ellis, EAPS; and Dr. Joyce Robinson.

Chapter 1. Introduction

James Clerk Maxwell: Mathematicians may flatter themselves that they possess new ideas which mere human language is as yet unable to express. Let them make the effort to express these ideas in appropriate words without the aid of symbols, and if they succeed they will not only lay us laymen under a lasting obligation, but, we venture to say, they will find themselves very much enlightened during the process.

The Boston Globe's "MIT 150"

On May 15, 2011, The Boston Globe published a special magazine recognizing 150 valuable contributions MIT has made in the worlds of technology, science, health care, culture, transportation, economics, and more. The Globe Staff Writers are Sam Allis, Hiawatha Bray, Scott Helman, and Carolyn Johnson, and the Globe Contributors are Scott Kirsner, Karen Weintraub, and Michael Blanding. The following are selections from the 150 contributions.

Number 1: World Wide Web Consortium: **Tim Berners-Lee**, the soft-spoken Briton who invented the Web in 1989 while working at a particle physics lab in Geneva, came to MIT in 1994 to help create the **World Wide Web Consortium**, to help spread technical standards for building websites, browsers, and devices (like televisions) that offer access to Web content.

Number 3: Transistor radio: Considered by some to be the most important invention of the 20th century, the **solid-state transistor** was born at Bell Labs in New Jersey. One of the three Nobel Prize laureate inventors was William Shockley, who earned his doctorate at MIT in 1936.

Number 6: The minicomputer: Ken Olsen and Harlan Anderson, who worked at MIT's Lincoln Laboratory in the 1950s, formed a start-up called Digital Equipment Corp. in 1957 to build what they called "**interactive minicomputers**," which would be smaller and less expensive than mainframes and designed with business use in mind.

Number 10: Where am I? : Like knowing where you're going? Ivan Getting founded the Aerospace Corp. A 1933 graduate, he designed radar systems at MIT's Radiation Laboratory during World War 2 and later worked at Raytheon Co. He was also one of the developers (and major advocates, in the face of Pentagon resistance) of a satellite-based **global positioning system** for navigation. You know it as **GPS**.

Number 16: Before Google: Perhaps Google owes a debt to MIT professor and administrator **Vannevar Bush**. Bush, who also served as a science adviser to President Franklin D. Roosevelt, conceived of a database-like device that would store all of an individual's books and correspondence and let that person search it instantly. The memex, wrote Bush, would be an "intimate supplement to his memory."

Number 19: Black box: Before World War 2, pilots couldn't figure out where they were without someone's help. Aviation pioneer **Charles "Doc" Draper**, who earned three degrees from MIT, insisted against prevailing wisdom that everything a plane needed to fly could be put into a "black box." Using a complex combination of spinning gyroscopes and sophisticated computer calculations, the MIT Instrumentation Laboratory he established created the first "**inertial guidance system**," flying from Hanscom Air Force Base to Los Angeles on the provenance of instruments alone. It was critical for airplane flight and for a precise piloting system for ballistic missiles, and it made the Apollo space missions possible.

Number 26: HP: **William Reddington Hewlett**, the cofounder of Hewlett-Packard Development Co., earned his master's in engineering from MIT in 1936.

Number 32: Drill, baby, drill: Back in the day, the techniques for finding new oil deposits were simple: Drill hole. If you didn't find anything, move over and repeat. In 1948, MIT researchers were among the first to apply modern scientific analysis to reports of seismic activity to create a chance above a "shot in the dark" in hitting pay dirt. Formalized as the **Geophysical Analysis Group** in 1952, the collaboration spurred the "digital revolution" in oil prospecting.

Number 34: Birth of the iPod... and more: As an MIT doctoral student in the early 1950s, **Robert Noyce** was known as "Rapid Robert" for his quick mind. As a cofounder of Intel Corp., he was known as the "Mayor of Silicon Valley." A mentor to Apple's Steve Jobs, Noyce is one of those credited as the inventor of the first practical integrated circuit, a.k.a. the "microchip" – which made modern-day computers possible. Without that, no iPod.

Number 41: The Bomb: Manson Benedict, an MIT graduate researcher during World War 2, discovered the process that ultimately led to the creation of **the atomic bombs** dropped on Japan and the peaceful application of nuclear energy for decades after.

Number 42: What's the "big" idea: **Alan Guth**, an MIT physicist, in 1981 came up with the inflation theory about the universe to answer puzzling questions created by the **big bang theory**.

Number 47: Calculated gifts: After amassing a fortune as a cofounder of Texas Instruments, with its ubiquitous calculators, **Cecil H. Green** made major gifts to institutions of higher education and medicine. Green, who earned two degrees from MIT in the 1920s, gave handsomely to his alma mater. At the time of his death in 2003, MIT said, Green and his wife, Ida, had given the equivalent of $91 million in current US dollars. **(The tallest MIT building is named for the Greens.)** One of his priorities was educating women; he funded a women's graduate residence hall and fellowships for women in science and engineering.

Number 48: Radar detectives: It is sometimes said the atomic bomb ended World War 2, but radar won the war. Much of that technology came from **MIT's Radiation Laboratory** – intentionally given a misleading moniker to divert attention away from the lab's true focus, on microwave **radar technology** development. Its contributions included airborne-bombing radar, a long-range navigation system, coastal defense radars, and early-warning radars. After the war ended, the Rad Lab closed. Seven years later, Lincoln Laboratory was founded, based on many of the same organizational principles and with many of the same employees.

Number 56: Way out there: The Apollo missions were tense times when the guidance, navigation, and control systems that were designed and developed at the **MIT Instrumentation Laboratory** underwent the ultimate test of guiding men to the moon. The lab, founded by **Charles Stark Draper** in 1932, was spun off from MIT in 1973 as Draper Laboratory. Draper's technologies have been important in missile guidance for submarine and land-based ballistic missiles and are now being used to tackle problems in biomedical research.

Number 62: Guessing games: MIT professor **Paul Samuelson** said that today's price of a given stock really is the best estimate of its actual value – and that money managers hunting for undervalued or overvalued securities are usually wrong. Samuelson was the first American to receive the Nobel Prize in Economics.

Number 68: **Whirlwind**: When the military wanted to build a flight simulator to train World War 2-era bomber crews, MIT got the contract. Whirlwind took a team of 175 people three years to build; when finished, it relied on about 4,500 vacuum tubes, and it was the first computer to use video displays (rather than paper printouts or flashing lights) for data output. It is considered the progenitor of the mainframes and minicomputers that businesses began to use in the 1960s.

Number 72: Taking off: Donald Douglas and James McDonnell, both MIT alums, cofounded **McDonnell Douglas** (which merged with Boeing in 1997), one of the world's biggest aerospace manufacturers.

Number 78: Stopping time: MIT professor **Harold "Doc" Edgerton** figured out how to marry a still camera and a strobe light that could flash up to 120 times a second, stopping time and allowing for the analysis of events too fast for the naked eye to see, like a bullet passing through an apple or a drop of milk splashing down. Using **stroboscopic**

photography, Edgerton's cameras also enabled the Allies to take nighttime reconnaissance photos from airplanes during World War 2 and photographed atomic bomb tests from miles away.

Number 83: That cool hologram: The "**rainbow hologram**" wouldn't exist if not for MIT professor Stephen Benton, who created one that can be seen in normal light.

Number 110: That's progress: **Robert Solow** is one of the world's leading economists and a 1987 Nobel recipient. His Solow-Swan growth model explained how technical progress and innovation are critical for any economy's productivity.

Number 119: All systems go: **Jay Forrester** grew up on a cattle ranch in Nebraska, worked on critical radar systems during World War 2, and helped design **computer technology that formed the basis for the US air defense system**. He drew on all these experiences to create a field at MIT in the 1950s called system dynamics, which became an influential way of analyzing just about everything that happens in the world. The idea was to use computer models and simulations not just for technical systems like aircraft, but to develop "management flight simulators" to help us learn about and manage complex human systems such as companies, markets, and cities.

Number 120: But can you spell it? : Penny Chisholm, an MIT biology professor, discovered a marine micro-organism called cyanobacterium Prochlorococcus. It's the most abundant microbe in the ocean. "Every fifth breath you take – thank Prochlorococcus for that **oxygen**," Chisholm has said.

Number 136: Like father, like son: **Theodore Miller Edison** was the fourth child of his famous inventor father. Ted Edison studied physics at MIT and after graduating in 1923 went on to earn more than 80 patents.

Wiener and the Geophysical Analysis Group

The Geophysical Analysis Group was based upon the mathematics of Professor Wiener who together with Professor Wadsworth oversaw the research. They gave me freedom to use my own discretion as to the path taken. Professor Wadsworth, an efficient organizer and eminent mathematical statistician, made available his extensive experience in operations research and weather forecasting. Professor Wiener, a kind and considerate gentleman, was always interested and enjoyed giving long discussions of his recent mathematical results as well as his ongoing research with colleagues in the biologic and medical sciences. I listened attentively. It was a great honor to be in his presence. Professor Robert Shrock (*Geology at MIT 1865-1965*, Volume 2, MIT Press, 1982) writes:

> The long-term result of the Geophysical Analysis Group (GAG) was the so-called "digital revolution" in exploration geophysics. The more immediate result, so far as MIT was concerned, was the philanthropy that produced the Green Center for the Earth Sciences, the three Green-endowed Professorships, McDermott Court with its impressive Calder stabile, the dozen annual McDermott scholarships, and numerous other less impressive but nonetheless important and helpful philanthropies described elsewhere in this history.

People today have become accustomed to life-changing advances in science and technology. Sometimes one has to look at the past in order to appreciate where we are and to anticipate what the future holds. In 1948, Prof. Norbert Wiener's published his book *Cybernetics*. In the book, Wiener foretold of the coming computer revolution. The term "cyber" is used now as a combining form meaning "computer" or "virtual reality." Examples are the words cybercrime and cyberspace. Wiener followed his book *Cybernetics* by his book *The Human Use of Human Beings*. There Wiener pointed out that the computer could be used for good or bad, and it was important that the good should win out. In 1942, Wiener had written a restricted report *Extrapolation, Interpolation, and Smoothing of Stationary Time Series with Engineering Applications* for the National Defense Research Council (NDRC). This report was republished as a book by the MIT Press in 1949.

If you wear eyeglasses, take off your glasses. Things become blurry. Put your glasses back on. Now things are clear and distinct. Your glasses act as a deconvolution mechanism for turning a blurry image into a distinct image. In other words, eyeglasses remove distortions in vision due to imperfections in the lenses of the eyes. In fact, everywhere you look, your glasses are constantly deconvolving the blurry image that you would otherwise see. If you wear glasses, you are constantly experiencing deconvolution.

The geophysicist needs to see the seismic reflections (i.e., echoes) from the subsurface layers of sedimentary rocks. The seismic records taken in oil exploration in 1950 looked blurry. A person looking at an eye chart might see the top rows (which have large-sized letters like a big E) but not the bottom rows (which have small-sized letters). In the same way, a geophysicist might see the large-sized reflections but not the small-sized reflections. Worst of all, he might not see any reflections at all. That would be a no-reflection (NR) case, which was all too prevalent in the days of long ago. Seismic "eyeglasses" were needed. In 1951 MIT was the only university that had a digital computer available to faculty and students. Whirlwind was to become the eyeglasses that would deconvolve the blurry seismic records to produce images with clear and distinct reflections. As a matter of speaking, Whirlwind made it possible for the geophysicist to see distinctly all of the letters on the eye-chart from top line to bottom line. Whirlwind gave the geophysicist 20/20 vision. Whirlwind did the deconvolution by the use of digital signal processing.

There are two components to sight: the eye, and the perception of the eyes' signals as processed by our brain. Accordingly, the act of seeing can be broken down into two steps. Step (1) Detection: When light enters our eye, the cornea and lens focus it onto a light-sensitive membrane called the retina at the back of the eyeball. The retina senses the received light and creates signals. Step (2) Imaging: The signals formed by the retina are dispatched through the optic nerve to the brain. The composition of these signals by the brain produces images that our conscious mind can interpret.

In contrast to human eyes, insects have compound eyes that bulge outward, allowing them to see in many directions simultaneously. As a result, an insect can see the direction of an enemy so it can take evasive action. The bulging convex eye is actually made up of thousands of mini-eyes (i.e., ommatidia). However, unlike human eyes, the eyes of insects tend to perceive shapes and outlines rather than crisp details. In summary: In a human, detection (without turning the head) has limited lateral range but imaging is precise. In an insect, detection (without turning the head) has extensive lateral range but imaging is imprecise.

Animals have very different eye structures. There is not just one refined eye structure found in all species of an animal group. The eyes of different species vary in their abilities to resolve images. However, in every case the eyes perform impressive feats for the particular animal's needs. Sophisticated eyes evolve because the eye has remarkable powers to help the animal to avoid predators, find food, and locate mates.

Let us now look at seismic exploration. It can be compared to an insect eye, not a human eye. Seismic exploration is like a compound eye looking into the earth. The earth has many layers. Light cannot penetrate into the earth, so light is of no help. Microwaves can penetrate a few meters deep, so we have ground-penetrating radar, which is of great value in waste disposal and archeology. Seismic waves can penetrate through the earth, so they are what we rely upon. When seismic waves are generated, the waves go through the earth and are reflected back. The reflected waves are recorded by many different receivers. Each receiver detects waves that have followed a particular travel path. Each receiver is like a specific lens. The compound eye of exploration is made up of all these receivers.

When an insect looks at a wall, it sees just the wall (i.e., one interface). The fictional character "superman" has the ability to see through solid objects. Seismic waves also have that ability. When a seismic wave travels into the earth it sees not just the first rock interface but all successive interfaces deep into the earth. That is fine, but unfortunately it also sees all the clutter (i.e., reverberations) between interfaces. Suppose that an unlucky mosquito has poor eyesight. Each mini-eye of the unfortunate mosquito needs to have its vision corrected. In other words we would have to fit a separate eyeglass to each of the thousands of its mini-eyes. It is the same in seismic exploration.

Step (1) Detection: Each seismic receiver has poor eyesight. It needs to have its vision corrected in order to eliminate the clutter. We have to fit a separate eyeglass to each of the thousands of receivers in a modern exploration survey. In other words, the signal recorded on each seismic receiver needs to be deconvolved. The source is systematically moved from place to place so as to cover the entire prospect. Each time that it is moved, all new eyeglasses have to be fitted.

Step (2) Imaging: After deconvolution and other such processes, the signals (now with eyesight corrected) must be imaged to yield a three-dimensional picture of the interior of the earth. These signals are dispatched, as it were, through the optic nerve to the brain (the computer), which composes the signals into images. The imaging, which is done by use of the wave equation, requires massive computer power.

Deconvolution is an important process. It can obtain a better estimate of the original form of a distorted signal. Deconvolution can often achieve amazingly good results. For example, in its original construction the Hubble Space Telescope had blurred vision. Once the Telescope was corrected by deconvolution, it produced the sharpest images of astronomical bodies ever obtained from the vicinity of Earth. Deconvolution can now be performed by a deconvolution chip. Such chips form the basis of cell phones and other communication devices.

The early work on deconvolution was done in 1950 and 1951 in the Mathematics Department under the guidance of Professors Norbert Wiener and George Wadsworth. In 1952 the work was transferred to the Department of Geology and Geophysics, now the Department of Earth, Atmospheric and Planetary Sciences (EAPS). The project was named the MIT Geophysical Analysis Group (GAG). The first step was the programming of Wiener's mathematics for the Whirlwind digital computer. The use of these programs on seismic data represented the first instance of digital signal processing. The GAG used Whirlwind not only for seismic deconvolution but also for other undertakings in geophysics. Digital geophysics began on Whirlwind.

During World War 2, the quickly constructed wooden MIT Building 20 was the home of the Radiation Laboratory, the place where radar was developed. The Geophysical Analysis Group was relegated to a few rooms in that venerable building. There were advantages to wooden buildings. You were allowed to nail things to a wall or knock out a wall. In the 1960s Professor Stephen Simpson boarded up windows so you would not know whether it was day or night when writing computer programs. It might have been an unconscious effort to suppress the circadian rhythm, which is a 24-hour cycle that regulates sleep and many other physiological processes. Electrical engineering research and biological research also found its home in Building 20. In time the wooden buildings at MIT were demolished one by one for modern structures. Building 20 was the last to go. It held many fond memories.

Of all the new MIT buildings, the Green Building is the most distinguished, the only tall building allowed on the MIT campus. It was the new home of the Department of Earth, Atmospheric and Planetary Sciences. MIT has always been the leader in digital geophysics. Generations of students have gone forth and blazed new trails. No matter how powerful computers become, it always seems that geophysicists want even more powerful ones. By the decade of the 1960s the digital revolution was well underway in geophysics. Digital geophysics was used in oil exploration and in the monitoring of underground nuclear explosions. In 1965 James Cooley and John Tukey published a paper on the fast Fourier transform describing how to perform it conveniently on a computer. This result prompted the electrical engineers and mechanical engineers to convert from analog to digital. At the present time computers are used in every act of human endeavor. Wiener's concern for the human use of human beings includes the human use of planet Earth and all its beings. We must be caretakers and caregivers. The past century was an era of discovery. The process is continuing. Now is also a time for rediscovery and preservation. The finite resources of the earth should be husbanded, not squandered.

Summary as given by R. R. Shrock

Professor Robert Shrock (*Geology at MIT 1865-1965*, Volume 2, MIT Press, 1982) writes:

> The great increase in demand for petroleum products during the years immediately following World War 2 presented the petroleum industry with the need for a greatly increased exploration program. As this program got under way, however, it soon became clear that the seismic methods developed in the 1930s and early 1940s were not sophisticated enough to explore successfully many of the potential oil-producing regions of the world. In particular the existing seismic methods were not very successful in exploring off-shore areas because of the water reverberations and in exploring deep strata because of the reverberations of the near-surface layers. By the early 1950s highly refined electrical engineering methods had been exploited almost to their fullest extent by the petroleum companies in attempts to filter the seismic data, yet these methods had failed to solve the reverberation problem except for the simplest cases. The situation was well described by Cecil H. Green of Texas Instruments Inc. when he stated, 'We have sharpened our tools about as much as we can; what we need now are some new tools.'

> What the MIT Geophysical Analysis Group (GAG) developed was just such a new tool. Whereas the existing seismic methods were analog in nature, the characteristic feature of the MIT GAG methods was that they were digital. These digital methods required for their successful use the accuracy, storage capacity, and speed of the large digital computer, of which MIT's WHIRLWIND with its magnetic core memory was the first true prototype.

The theoretical work of the GAG involved the linkage of the mathematical work of Professor Norbert Wiener with the physical theory of seismic wave propagation: namely, that the phase-characteristic inherent in the design of a Wiener digital filter is in fact the same phase-characteristic that the earth imposes on a propagating seismic wave. The empirical work of the MIT GAG involved showing that these digital methods could automatically transform, within the computer, a field record that could not be interpreted by existing methods into a record that would yield the required information about the subsurface structure.

Owing to a decline in exploration activity in the late 1950s, due largely to the great Middle East discoveries, the use of these digital methods did not become widespread until the 1960s; the essentially complete conversion of seismic exploration at that time, from analog to digital methods, constitutes the so-called 'digital revolution' in geophysical exploration. Since the mid-1960s virtually all seismic exploration for petroleum has made use of the digital methods developed by the MIT GAG. The discoveries of off-shore oil and natural gas deposits as well as many of the deep oil fields on land made in the last decade represent the fruition of the original research pioneered by the small group of MIT professors, their graduate students, and the industrial representatives. And the work of these few is carried on today by the billion dollar industry of seismic exploration for petroleum, an industry that is one of the leading users of digital computers.

In summary, then, it can be said that the MIT GAG Project produced at least three major impacts on the earth sciences at MIT. First, it was the successful application of Wiener's mathematical work to the analysis of seismic records--an accomplishment that greatly aided the 'digital revolution' in exploration geophysics. Second, it provided a unique opportunity for a score of able students to get prepared for a future career. Third, it helped to motivate the philanthropies of the Greens and McDermotts that have been of inestimable value to the advancement of the earth sciences at MIT. Finally, the MIT GAG Project is a dramatic example of what can be done when academic and industrial professionals work closely together for a common purpose."

MINUTES OF FIRST ANNUAL MEETING OF THE ADVISORY COMMITTEE FOR MIT'S

GEOPHYSICAL ANALYSIS RESEARCH PROJECT   Cambridge, Massachusetts

August 12 - 13, 1953

EXPLANATORY NOTES: (1) The attached mailing list gives the 14 companies participating in the project.   The list of designated recipients for these minutes is given in the second column and the list of those attending in the third.   Copies are being sent to both lists.   Please, notify the secretary (W. J. Yost) of any correction to this list. Some additional copies are available if desired.

(2) Because of the nature of this meeting, the minutes are not presented in chronological order.   Instead, an attempt has been made to group these notes according to subject matter.

I.   BUSINESS SESSIONS

Mr. Silverman presided and Mr. Hurley presented the administrative information for the MIT group.   A financial summary for the five months (2/1/53 - 7/1/53) was given and discussed in terms of 5/12 of the original budget estimate.

5/12 of Budget ($32,000)		$13,333.33
Salaries	$6,505.99	
Raytheon Computing Contract	9,180.00	
Miscellaneous	1,650.00	
Total Expenses	17,335.99	
Deficit 7/1/53		$17,335.99
		$ 4,002.66

Most of the deficit ($3,300.00) represents computations under contract, but not performed during this period.   Miscellaneous includes cost of reproducing reports, travel by project personnel and operational supplies and materials.   It was emphasized that the computations on MIT's Whirlwind computer would have cost $10,500, if paid for at the customary rate.   Also $7,500 for similar computations prior to 2/1/53 could be considered as applying to this project.

The proposed budget for the year 7/1/53 - 6/30/54 was presented on the basis of $56,000 contributed.   Mr. Hurley presented this income as $60,800 less $4,800 for MIT overhead charges or $56,000. He indicated that only one company (of the group not in the Industrial Liaison Program had paid the $800 for overhead charges.   He was not certain whether the additional 4,000 would be received or whether it would come out of the

Chapter 2. Analog versus digital

Gottfried Leibniz: Music is the pleasure the human mind experiences from counting without being aware that it is counting.

Industrial revolution and digital revolution

The Industrial Revolution began in the middle of the eighteenth century. It marked the transition to new manufacturing processes that eliminated much of the hand labor. It involved the invention of machines for manufacturing, the use of steam power, and the development of factories. The Industrial Revolution was a turning point in history; it affected every aspect of the way that people lived. In the industrial revolution, science and technology were characterized by technical drawings, blueprints, and working models. In fact such physical evidence had to be offered in order to obtain a patent.

The Digital Revolution marked the transition to digital processes that could eliminate various forms of tedious labor. It began with the invention of the ENIAC, the first high-speed electronic digital computer which came on stream in 1946. A plethora of new digital computers evolved that were responsible for the creation of a knowledge-based society. Computers influence communication and control at every level. The Digital Revolution is a turning point whose development is still underway. Today the digital computer touches every aspect of daily life. In the digital revolution science and technology are often characterized by mathematics, and especially mathematics that can be represented by numerical quantities.

Dean Clark (*The Leading Edge*, Society of Exploration Geophysicists (SEG), Tulsa, Oklahoma, June 2005, p. 572) writes:

The word "revolution" is dramatically overused in contemporary journalism but there is no doubt that it is doubly apropos for geophysics in the 1960s which underwent two major and simultaneous "back to basics" overhauls in that decade (not counting the social revolution for which that decade is best remembered by the general public).

The geophysical revolutions were technological and theoretical. The former, the switch from analog to digital instrumentation, had been brewing since the early 1950s when MIT's Geophysical Analysis Group (GAG) demonstrated that statistical analysis (via computer manipulation of data recorded as time series) could dramatically improve the quality of seismic reflection records. The latter, the emergence of plate tectonics as the new paradigm for explaining the inner workings of planet earth, struck with a tsunami-like suddenness that blindsided nearly all of the world's experts.

The digital revolution completely changed the workflow of exploration geophysics. The analog era was based on field records. There was essentially no data processing, and the mathematical computations (by a crew member with the title of "computer") to clean up the data were minimal. In the digital era, the purpose of the field record is (to be blunt) merely to verify that somebody hit the "on" button on the equipment. It is hardly ever looked at again except, perhaps, as a historical curiosity.

The digital revolution would sooner or later, usually the former, transform every technology-based discipline and virtually every aspect of routine daily existence in countries with "advanced" economies. Exploration geophysics, however, had gotten a head start from MIT graduate student Enders Robinson who laid the foundation for the Analog-to-Digital (A/D) metamorphosis en route to receiving his PhD in 1954. Many, many other individuals obviously made significant contributions that facilitated the quick onset of this new era in exploration geophysics but two that will be cited here are Ken Burg and Jack Kilby.

Burg was a first-generation geophysicist whose career began in 1927—just in time to eyewitness a previous technological revolution, when reflection began to supplant refraction as the dominant seismic technique. Burg became the research director at Geophysical Service Incorporated (GSI) in 1947 and he was the personification of GSI's "revolutionary not evolutionary" research philosophy. Primarily because of Burg's enthusiasm, GSI's leadership committed the company to developing digital technology in the 1950s when the rest of the industry, despite the Geophysical Analysis Group results, decided to focus on refining the existing analog methods.

Kilby was the engineer at Texas Instruments who, in 1958, invented the integrated circuit (a.k.a. computer chip) which accelerated the evolution of computing technology into the exponential growth pattern that has now reigned for nearly half a century. At that time, Texas Instruments was a subsidiary of GSI, giving the latter supreme (and justified) optimism about the future of computing and the cash flow to make the huge financial commitment required to develop digital technology from ground zero. Ironically, in a "real-life" reenactment of a Greek tragedy, Texas Instruments would end up killing its parent, GSI, a few decades later, but that's another story. [Jack Kilby was awarded the 2000 Nobel Prize for Physics for his work on the integrated circuit.]

Most leading geophysical contractors were advertising digital services in the journal GEOPHYSICS by the summer of 1965. The company *Independent Exploration* announced its new digital recording system in February 1966 with a type size that most contemporary newspapers were reserving for the start of World War 3. Numbers tell the rest of the story. SEG's annual survey of geophysical activity reported that 333 crew months of digital data were recorded in 1964—about 4% of the total. The number grew to 14% in 1965, 34% in 1966, 82% in 1972, and 91% in 1973 (the last year, because total saturation seemed inevitable, statistics were collected concerning instrumentation). The analog equipment was soon banished as ruthlessly as the torsion balance had been 30 years previously. As Carl Savit said in 1986, "Even if we wanted to recreate the old analog gadgets, I doubt the skill to build them is here anymore."

More basic: analog or digital?

Analog refers to continuous quantities. Digital refers to discrete quantities. An analog watch is one that shows the time by rotating hands. A digital watch is one that shows the time by displayed digits. One gets the impression that an analog clock indicates every possible time of day. In actuality the hands do not rotate smoothly, but in series of jumps. For example, the second hand might jump at every tenth of a second. An inexpensive analog alarm clock would be a low resolution analog device. An expensive Swiss analog watch would be a high resolution analog device.

Which is more basic—analog or digital? In other words, is the world continuous or is the world discrete? People experience the world in an analog manner. Vision, for example, is an analog experience because we perceive infinitely smooth gradations of shapes and colors. Color is how our eyes and brain interpret light. In this sense, colors are continuous. However, every person's eyes perceive color a bit differently. It is estimated that the average person can see several million distinct colors. In this sense, colors are discrete.

The world has experienced a digital revolution. Digital is an abrupt departure from analog. The word digital is an adjective meaning "of or relating to a finger or fingers." Mankind first counted with fingers, and thereby we obtained the ten discrete digits 0,1,2,3,4,5,6,7,8,9. The word digital also refers to signals or data expressed as series of numbers, typically represented by values of a physical quantity such as voltage or magnetic polarization. In photography a film camera is considered as an analog device. The film represents the changing values of continuously variable physical quantities. There are smooth gradations of shapes and colors. The term "analog" differentiates such cameras from the digital cameras. An image taken by a digital camera is recorded as a matrix of numbers. Analog television encodes television and transports the picture and sound information as analog signals. Such signals are formed by varying the amplitude and/or frequencies of the broadcast signal. A digital television transports the picture as digital signals. Such signals are formed by varying the numerical values of the broadcast signal. There are notable non-electrical analog devices, such as various clocks (such as sundials, water clocks, pendulum clocks, mechanical wristwatches), the astrolabe, slide rules, the governor of a steam engine, the planimeter (a simple device that measures the area of a closed shape), Kelvin's mechanical tide predictor, acoustic rangefinders, mercury thermometers, bathroom scales, and the speedometer of a car.

An analog device uses one physical quantity to represent another physical system. An example is the wind tunnel. A wind tunnel is an analog device used to study the effects of air moving past solid objects. A wind tunnel consists of a tubular passage with the object under test mounted in the middle. Air is made to move past the object by a powerful fan system. The model is equipped with sensors to measure aerodynamic pressure and other variables. Wind tunnels are used to develop airplanes. Instead of the air standing still and an airplane moving through it, the same effect is obtained by letting a model airplane stand still and the air move past it. In that way a stationary observer can study the flying object in action, and can measure the aerodynamic forces being imposed on it. Wind-tunnels are also used to determine ways to reduce the power required to move automobiles at various speeds.

An analog signal is a continuous signal that represents a time varying quantity. Some examples of analog signals are electrical signals, sound waves, water waves, seismic waves, light waves, microwaves, and radio waves An analog signal differs from a digital signal, which is a sequence of discrete numbers. An analog signal is subject to noise and distortion introduced by communication channels which can progressively degrade the signal. Converting an analog signal to digital form introduces quantization error. However once in digital form the signal can in general be processed or transmitted without introducing additional noise or distortion. As analog systems become more complex, they can degrade signal resolution markedly. This explains the widespread use of digital signals in preference to analog. In digital systems, degradation can not only be detected but corrected as well.

Most of the analog entities can be simulated digitally. Photographs in newspapers, for instance, consist of an array of dots that are either black or white. From afar, the viewer does not see the dots (the digital form) but only lines and shading, which appear to be continuous. From such an example, we might think that digital representations are approximations of analog events. However, the opposite is true at the most basic level of physics, the micro-world.

Electromagnetic waves have discrete energy packets (called quanta). The quanta behave in a manner similar to that of particles. This phenomenon is the basis of the branch of physics known as quantum mechanics. Quantum theory generalizes all classical theories in physics except general relativity. Quantum theory is useful because it provides accurate descriptions for many previously unexplained phenomena. At this most basic level, the analog energy that humans perceive is only an approximation of the true digital energy that governs the quantum-world.

Nature and numbers

What is analog technology? What is digital technology? Analog has to do with comparisons of natural things. Digital has to do with manipulation of numbers. A cook measures quantities of various ingredients by comparison to fixed weights by means a balance or any of various other instruments for weighing. For volumes a cook uses a measuring cup. We experience the food with our senses. An accountant counts money and writes down the amount. The amount is a number with two decimal places, such as $103.26. We experience the number with our intellect.

An analog watch displays time with hands that rotate around the dial. The location of the hands gives a measurement of the time. The movement of the hands is an analog of the natural passing of time in daily cycles. We experience analog time with our senses. A digital watch give the time in numbers for hours, minutes, seconds, as 9:58:24 . We experience the numerical time with our minds. Instead of measuring time from rotating hands on an analog watch, we simply read the numbers shown on the display of a digital watch.

If you place a wooden or plastic ruler on your finger, the scale on the ruler gives a measurement of the length of your finger. You have to read the scale to obtain the numerical length. You put a mercury thermometer into your mouth. The level of mercury on the thermometer is a measure of your temperature. In other words, the level of mercury is an analog of your temperature. You have to read the scale on the column of mercury in order to obtain a numerical value of your temperature. A digital thermometer directly gives your temperature as a number.

Until the present digital age, almost every instrument was analog. A moving-coil meter with a pointer was used to measure an electric current. The more the pointer moved, the higher the current in the electric circuit. Generally speaking, a device is analog if it does not handle information by the processing of numbers. A film camera is an analog technology. The image is captured on film coated with a silver-based compound. When developed the film gives an image. The image is an analogy of the thing you photographed. In digital photography there is no need for film, chemicals or darkrooms. Arrays of photo sensors capture images as a collection of numbers. The numbers are processed by computer software.

In analog devices, we store words, pictures, and sounds as representations on things like plastic film or magnetic tape. In digital devices, we first convert the information into numbers and store or display the numbers. An inkjet printer prints the numerical representation of a picture into analog form by firing tiny jets of colored ink or dyes at paper. Language is made up of words. The word "dog" is a representation of the animal dog. The word "barks" is a representation of a dog making a sound. Language tends to be analog. Mathematics is made up of numbers and symbols. A line is described as the shortest distance between two points. A mathematical point is infinitesimally small. We cannot see a mathematical point. A mathematical line is infinitesimally thin. We cannot see a mathematical line. However analog visualizations of both point and line are perfectly clear in our mind's eye.

Pure mathematics is mathematics studied in its own right. Broadly speaking, pure mathematics deals with abstract concepts. Applied mathematics is mathematics applied to other disciplines such as navigation, astronomy, physics, economics, engineering, biology and medicine. Applied mathematics deals with physical things. The electronic digits that are rushing around inside a digital computer are about as mathematical as you can get. At the input to the computer you generally need an analog-to-digital converter to enter the reality of the analog world into the digital computer. At the output from the computer you generally need a digital-to-analog converter to recapture the reality of the analog world.

On a check the amount is written in two ways. The first way is digital (numbers), as $405.06. The second way is analog (words), as "four hundred five dollars and six cents." Shakespeare's Macbeth says: "Nothing is but what is not." In the analog way, **nothing** is "**no thing**," so the word zero is not needed. A Roman numeral is analog, with letters used as abbreviations for words (as X for ten, as C for hundred). Because zero is not used in a Roman numeral, the Romans had no symbol for zero. In due time, the Western world introduced the **symbol 0 for zero** and switched from the analog Roman numerals to digital numerals. This first digital revolution was completed by Shakespeare's time. King Lear says, "Nothing will come of nothing." The word "nothing" appears again when the Fool tells Lear he is nothing without his crown and power: "Now thou art an O without a figure [zero]. I am better than thou art now. I am a Fool [analog], thou art nothing [digital]."

Applied mathematics tries to bridge the gap between thing (analog) and symbol (digital). At MIT, Professor Paul Samuelson and Professor Robert Solow applied mathematics to economics. Professor Norbert Wiener applied mathematics to the study of communication and control in the animal and machine, and came up with the field of cybernetics. Professor George Wadsworth applied mathematics to the study of weather prediction. Professor George Valley used mathematics to build a wide-ranging continental radar network (SAGE) with control and communication provided by an assembly of advanced computers based upon Whirlwind.

Increasingly common today is a side effect of something called digital motion sickness or cyber-sickness. It causes a person to feel woozy, as if on a boat in a churning sea, from viewing moving digital content. The sense of balance is different from other senses in that it has many inputs. When inputs do not agree, a feeling of dizziness and nausea can result. In traditional motion sickness, mismatch occurs. Movement is felt in muscles, joints and inner ear, but it is not seen. That is why getting up on the deck of a ship and looking at the horizon helps a seasick person to feel better. However with digital motion sickness, it is the opposite. Movements like turns and twists are seen on screen but are not felt. The result is the same. A person can have sensory conflict that gives a queasy feeling.

Examples of analog and digital

The names of teeth originally were analog. Now they are digital. For example, the Wisdom Tooth (3rd Molar) in the upper right quadrant is number 1 and the second Bicuspid in the lower left quadrant is number 20. The names of golf clubs originally were analog. Woods are long-distance clubs used to drive the ball down the fairway. Woods have a large head and a long shaft for maximum club speed. Historically woods were made from wood. Irons are clubs with a solid, all-metal head featuring a flat angled face. They have a shorter shaft and more upright lie-angle than a wood. The practice of giving numbers to golf clubs began in the 1920s, as industrial production became prevalent. The analog names were replaced by numbers. For example, the mid-iron became the 2-iron, the mashie became the 5-iron, the lofting-iron became the 8-iron and the niblick became the 9-iron.

Up until the present digital age, analog clocks have been used to measure and keep track of time. A sundial is indeed a true analog clock for the sun moves continuously. The gnomon is the part of a sundial that casts the shadow. At an ancient Egyptian temple, an obelisk served as a gnomon. The moving shadow cast by the gnomon was measured by markers. The position of the shadow allowed the Egyptians to calculate the time of day. However, because of the earth's tilt, the sun's path through the sky changes slightly from day to day, so the shadows cast by the gnomon are not the same every day. Many sundials overcome this problem by angling the gnomon and aiming it north. This type of gnomon is called a style. Because its alignment compensates for the Earth's tilt, the hour marks remain the same all year round.

A digital clock is capable of representing only discrete time instants — for example, at an increment of every tenth of a second. However, by making the time increment smaller and smaller, digital clocks can be made that outperform any analog clock. The smallness of the increment is the key to any digital system. If the temporal and spatial increments are small enough, a digital system always can outperform the corresponding analog system.

The ancient Egyptians created the water clock (a.k.a. clepsydra) and the Greeks perfected it. Water clocks were among the earliest timekeepers that did not depend on the observation of celestial bodies. The water clock is an automatic self-regulatory device. A water clock uses a vessel with a hole near the bottom that allows a stream of water to escape. In order to use the escaping water as a measure of time, it is necessary to keep the escaping stream flowing at a constant rate. The water flows at a constant rate only if the level of water in the vessel is held constant. A continuous flow of water into the top of the vessel is used. This flow must be greater than the loss of water at the bottom of the vessel. An overflow pipe is supplied near the top of the vessel to fix the height of the water. Thus a constant water level is maintained in the reservoir as required. The ancient Greek philosopher Plato (428 BC–348 BC) was said to possess a large water clock with an alarm signal similar to the sound of a water organ.

The first mechanical clocks employing the verge escapement mechanism were invented in Europe at around the start of the 14th century. They were the best clocks until the pendulum clock was invented by Christiaan Huygens in 1656. A pendulum clock uses a pendulum (i.e., a swinging weight) as its timekeeping element. The advantage of a pendulum for timekeeping is that it is a close approximation to a harmonic oscillator. It swings back and forth in an almost precise time interval dependent on its length, and it resists swinging at other rates. Until the 1930s, the pendulum clock was the world's most precise timekeeper.

Attendant to the age of enlightenment in the seventeenth century is the age of technology. The age of technology is characterized by the use of higher mathematics to design instruments. Certainly Leonardo was a great inventor. His designs foretold a plethora of later inventions, but he used no higher mathematics. Christiaan Huygens is the father of technology. His 1656 book on his invention of the pendulum clock is the first instance of the use of higher mathematics to develop an instrument. Specifically Huygens used a mathematical proof to show that a pendulum does not swing according to simple harmonic motion. Huygens foresaw calculus. He taught mathematics to Gottfried Leibniz, the co-inventor of calculus with Newton.

Image-making

An image is a reproduction of the form of someone or something. Persons often tend to think in terms of images. Most important are the "memory pictures" in our minds. Our ability to deal with these images in different contexts leads to formation of ideas and concepts. These entities seem to serve as elements of thought when they are reproduced, combined and analyzed. This mental imaging seems to be the essential feature of productive thought before any connections with words and other signs are made.

More tangible are the images drawn on some surface. The first such images are the exquisite paintings of animals made on cave walls by prehistoric people. An image cannot be an exact reproduction of the object in question, but must contain the essential features pertinent to the task at hand and the artistic nature of the work. The first imaging system was the artist, and the technology was that of his paints and materials.

One of the most pressing tasks in the development of civilization was the requirement of mankind to make permanent images of where we stand in relationship to the things around us and to the universe in general. Always associated with image-making has been geometry, and mathematics in general. Practical geometry was used by ancient people in such enterprises as building houses and laying out land divisions. In fact, the word "geometry," which is of Greek origin, means land measuring. From the early beginnings image-making has progressed to the high-technology systems in use today.

Ancient peoples were eager to understand the world, but lacked the image-making skills required to make an accurate picture. However, by the fifth century BC, Greek scholars determined that the Earth was a globe. Over the ensuing years, the Greeks made remarkable advances in mathematical geography. Even though they did not have enough observed detail about the surface of the Earth to make a map of the whole world, they were able to make some surprisingly accurate estimates based on mathematics and astronomy. In this way, the ancient Greek civilization developed a workable image of our place in the universe. They determined that the sun and moon were spherical bodies each at a fixed distance from the Earth. Aristarchus of Samos (ca 310 BC-ca 250 BC) asserted that the sun was the center of the universe, and that the Earth revolved around the sun. He was able to give values for the sizes and distances of the sun and moon and these values, although approximate, were determined by mathematically correct principles. Eratosthenes (ca 276 BC-ca 198 BC) was first to estimate the circumference of the Earth.

To obtain the best possible images, we must turn to digital methods. The Cloisters is a museum of medieval art in upper Manhattan. It contains seven beautiful tapestries known as "The Hunt of the Unicorn." The tapestries are twelve feet tall and up to fourteen feet wide, woven from threads of dyed wool and silk, some of them gilded or wrapped in silver. In order to preserve a record of the colors and the mirror images, the Unicorn tapestries were photographed in a series of individual digital photographs which filled more than two hundred compact disks. Professors David Chudnovsky and Gregory Chudnovsky fed the data into a computer and joined the individual photographs together into seamless images of the tapestries. The result was the largest and most complex digital image of any art work ever made. The digital images can be readily duplicated and distributed to other museums. The museums can then print the digital image for people to see. Such replication helps to safeguard the art work over the vicissitudes of time.

Honey bee

Isaac Watts (1674–1748) writes:

> How doth the little busy bee
> Improve each shining hour,
> And gather honey all the day
> From every opening flower.
>
> How skillfully she builds her cell;
> How neat she spreads her wax,
> And labors hard to store it well
> With the sweet food she makes.

A fascinating example of visual thinking is found in the action of a forager bee. When a forager bee discovers a new source of flowers with pollen or nectar, she must convey the location of the new source to the other forager bees. The bee returns to the hive and then dances at a particular region in the comb. This region is called the dance floor. Honey combs are usually vertical, so the dance would be performed on a vertical plane. The vertical dancing floor is inside the hive and thus quite dark. The bee dance must provide information of horizontal directions on a vertical plane.

Instead of looking at a map in a vertical position, a person lays it down on a horizontal plane with the north arrow on the map pointing in the direction of due north on the ground. In this way the horizontal directions on the map coincide with the corresponding horizontal directions on the ground. However, bees cannot do that. Bees are quite unwilling to topple over their beehive so that it lies flat on the ground. Instead, a forager bee uses the sun compass; that is, the bee remembers the angle that her path makes with the direction to the sun. For example, if she were flying directly toward the sun, the angle would be zero degrees. If she were flying directly away from then sun the angle would be 180 degrees. For example, suppose the direction to the source of pollen is 286 degrees by the sun compass. The question is how would the forager bee communicate this information to her sister bees when she arrives back at the hive?

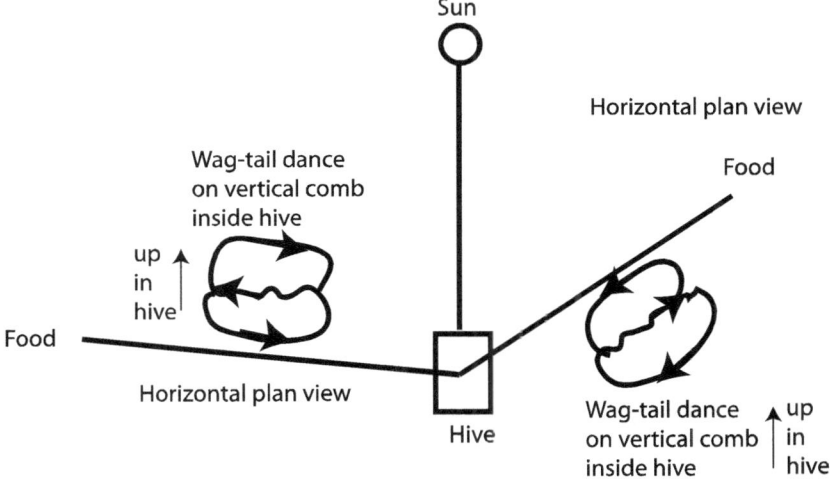

Figure 1. The tail-wagging dance

The forager bee does the tail-wagging dance (Figure 1), which is a figure-eight dance particular to the honeybee. By performing this dance, the forager tells her sisters the direction and distance to patches of flowers yielding nectar and pollen. More exactly, the tail-wagging dance communicates both the distance and the direction to the discovered source.

Here is a description of the dance. On the dance floor of the hive, the forager bee runs straight ahead for a short distance, returns in a semicircle to the starting point, runs again through the same straight course, then makes a semicircle in the opposite direction to complete a full figure-eight circuit. The time duration of the straight course gives the distance to the nectar. The conversion factor is 0.65 miles per second. For example, if the straight run lasts 2 seconds, then the pollen source is 1.3 miles away. In other words, bees measure distance in terms of time. It is remarkable, that a bee can use a thinking process that is conscious of the passage of time. The bee, like people, can measure distance in terms of time. For example, Newburyport is 40 minutes away from Boston by car. New York is 5 hours away from Boston by train.

Let us get back to the business of the bees. A forager bee can cover a distance of up to 10 miles from the hive. The orientation of the dancing bee during the straight portion of her tail-wagging dance indicates the location of the food according to the sun compass. The forager bee must communicate horizontal directions on a vertical plane, which is usually in the dark. The natural reference in a vertical plane is gravity, so the dancing bee replaces the real reference, the sun, by the up-direction. The bee transposes the solar angle into a gravitational angle. For example, if the path of the forager bee were a straight line at 270 degrees with respect to the direction of the sun, then the straight portion of her dance would be 270 degrees with respect to the upwards vertical direction.

All well and good, but a bee would have to know that the sun continually changes its position in the sky by one degree every four minutes. In fact, the bee automatically and continuously makes this correction. In other words, in travelling to the flowers, the bee orients itself by its angle with respect to the sun taking into account the movement of the sun from east to west. When the bee cannot directly see the sun, she will infer the sun's position from the polarization pattern of the sky. They can maintain course under the shadow of a mountain, at twilight, or when only small patches of sky are not clouded. A forager bee who returns too late in the day waits till morning to do the dance. This forager not only remembers the direction and distance, but also corrects for the direction of the morning sun.

MASSACHUSETTS INSTITUTE OF TECHNOLOGY
CAMBRIDGE, MASS.

E. A. Robinson

Received

Action

P. M. H.

Professor P. M. Hurley
24-416

        I have looked over the records which Dr. Bradley of Cities Service sent. The traces on these records are very close together, and the records are photostatic re- productions, some of which are not too clear. These factors make the reading of these records difficult at best. Nevertheless, they form an interesting set of records to analyze. From the records we can probably read enough traces to give an indication of the applica- bility of our method.

EAR | bh

Chapter 3. Remote Sensing

Max Planck: The entire world we apprehend through our senses
is no more than a tiny fragment in the vastness of Nature.

Methods of perception

Senses refer to physiological methods of perception. Aristotle listed five
senses: sight, hearing, touch, smell, taste. In addition, people have other
senses, which include a sense of balance, a sense of time, a sense of
motion and acceleration, a sense of temperature differences, and a
sense of pain. Certain animals have senses that humans do not possess.
Spiders can detect mechanical strain in the exoskeleton, providing
information on force and vibrations.

Several species of fish, sharks, and rays have the capacity to sense
changes in electric fields in their immediate vicinity. Other fish actively
transmit their own weak electric signals and sense the pattern of
returned electric signals over the surface of their bodies. Among
mammals, the platypus has the most acute sense of fluctuations in
electric fields.

Animals have various ways of navigating. Some birds and insects such
as bees have the ability to detect fluctuations in magnetic fields.
Although this sense is not a well understood, it is essential to the
navigational abilities of migratory birds. Homing pigeons can use
magnetic fields as part of their complex navigation system. Homing
pigeons orient themselves correctly on a clear, sunny day by using the
movement of the sun during the day. However, pigeons on overcast
days also navigate correctly. It seems that homing pigeons can rely on
magnetic fields to orient themselves.

Migrating birds can cover thousands of miles in their annual travels,
often traveling the same course year after year with little deviation.
First-year birds often make their very first migration on their own.
Somehow they can find their winter home despite never having seen it
before, and return the following spring to where they were born.

The secrets of their amazing navigational skills are not fully understood, partly because birds combine several different types of senses when they navigate. Birds can get compass information from the sun, the stars, and by sensing the earth's magnetic field. They also get information from the position of the setting sun and from landmarks seen during the day. There is evidence that the sense of smell plays a role, at least for homing pigeons.

Some species, particularly waterfowl and cranes, follow preferred pathways on their annual migrations. These pathways are often related to important stopover locations that provide food supplies critical to the birds' survival. Smaller birds tend to migrate in broad fronts across the landscape. Many small birds take different routes in spring and fall, to take advantage of seasonal patterns in weather and food.

Microscope

The microscope was developed concurrently with the telescope. Robert Hooke (1635-1703) and Anton van Leeuwenhoek (1632-1723) were leading figures in this development. It is not known who first invented the microscope, but as early as 1625 the physician John Faba (1574-1629) introduced the name "microscope" because it permits a view of minute things. In 1665 Hooke published his *Micrographia* which included a description of the microscope and its uses. Hooke's illustration of small insects and organisms were so beautiful that they were used in textbooks for over two hundred years. Leeuwenhoek was the first to discover microorganisms (a.k.a. microbes), and his work opened up the fields of microbiology, entomology, and crystallography. Microscopic technology has advanced greatly since the days of these pioneers. The 20th century development of non-optical microscopy, such as the electron microscope, has increased our image-making capabilities to the point of where we can see things on an atomic scale.

Two events in the 1920's led to the development of the electron microscope. One was the idea first advanced by de Broglie in 1924 that particles have wave properties. In particular, very short wavelengths are associated with an electron beam. The other event was the discovery made by Busch in 1926 and 1927 that a suitably shaped magnetic field can be used as a lens. The development of these concepts brought about the design and construction of a practical electron microscope as early as 1938. Since then there has been a continuous development of this technology.

One of the most important practical problems is in the obtaining of specimens thin enough to allow sufficient transmission of electrons. The transmitted electrons need to affect the photographic material satisfactorily but not to affect the specimen detrimentally, as by the heat of absorption of electrons. Another problem is the interpretation of images, a problem with which people have struggled for centuries.

Instead of transmitting the primary electrons to form an image, the scanning electron microscope (SEM) utilizes the secondary electrons reflected from the surface of a specimen to produce an image of the topography. In such a reflection microscope the contrast between features is often too low, but it may be enhanced by signal processing. The computer has significantly enhanced the image-making process by increasing the depth of focus.

Microbes as an analog device

The entire microbe ensemble in the biosphere may be considered as an analog device. A microorganism (or microbe) is a microscopic living organism, which may be single celled or multicellular. Microbiology is the study of microorganisms. Microbes are diverse and include all the bacteria and archaea and almost all the protozoa. They also include some fungi, algae, and certain animals, such as rotifers. Viruses are often classified as microorganisms, despite the fact they are considered as nonliving. Without microbes, life on Earth as we know it would not be possible.

Microbes are found in many places on earth. They live in every part of the biosphere, including soil, hot springs, "seven miles deep" in the ocean, 40 miles high in the atmosphere, and inside rocks down within the Earth's crust. Microbes, under certain test conditions, have been observed to live in the vacuum of outer space. Microbes are adaptable to all sorts of conditions, and survive wherever they are. They thrive in the Mariana Trench, which is the deepest part in the oceans. They thrive inside rocks deep below the sea floor. They thrive hundreds of feet within the Antarctic ice. Microbes play a crucial role in the Earth's ecology. They decompose substances. They are a vital part of the nitrogen cycle. Airborne microbes play a role in the weather. Microbes are increasingly used in biotechnology. Some microorganisms are pathogenic and cause disease and even death in plants and animals.

Microbes were the first forms of life to develop on Earth. They appeared on Earth approximately 3 to 4 billion years ago. True multicellular organisms that solve the problem of regenerating a whole organism from germ cells appeared much later. The vertebrates have only been around for the past half billon years. Microbes tend to have a relatively fast rate of evolution. Most of the microbes reproduce rapidly. Bacteria are also able to freely exchange genes through conjugation, transformation and transduction, even between widely divergent species. This horizontal gene transfer, coupled with a high mutation rate, allows microbes quickly to learn to survive in new environments. This ability has led to the development of strains of bacteria resistant to medical antibiotics.

Microbes can be found throughout the ocean, from rocks and sediments beneath the seafloor, across the vast stretches of open water, to intertidal and surf zones. The diversity and number of microbes in the ocean far exceed that of macroscopic life, and many employ unique life strategies not seen anywhere else on Earth. Some microbes are photosynthetic, deriving their energy from the sun. At deep-ocean sites and around hydrothermal vents and seeps, microbes derive energy from chemical reactions and not the sun. Some microbes prey on others; some obtain carbon from inorganic sources; some are scavengers that feed on dead organisms or other waste organic matter. Certain microbes in the seas can eat hydrocarbons, such as naturally seeping petroleum and petroleum from spills. In other words, fossil life (petroleum) feeds present life (bacteria and animals that eat bacteria).

Ocean microbes play an important role in Earth's biogeochemical cycles, particularly the carbon, nitrogen, phosphorus, iron, and sulfur cycles. They also form the very base of the marine food chain, recycle nutrients and organic matter, and produce things needed by higher organisms to grow and survive. Many have evolved over millions of years to form symbiotic relationships with animals that allow the host animals to live in harsh or otherwise toxic environments, such as the hot, sulfur-rich waters around hydrothermal vents.

Microbiology is the study of microscopic organisms. It is true that certain microbes are associated with various sicknesses. However, most microbes are responsible for beneficial processes. Microbes demonstrate the cognitive abilities of decision-making, group communication, and group behavior. Microbes can communicate with each other in order to form structures and function as a multicellular creature. Microbes use an elaborate language of signals which elicit a wide range of other behaviors. Some microbes use secreted chemicals as messages. One chemical message tells others that there is a lack of food in a particular location. The microbes that receive this message then go in other directions.

Individual microbes send out signals that communicate their presence, and when a certain number have signaled they launch various group activities. This is called "quorum sensing." For example, some colonies of bacteria light up when enough bacteria are present. Similarly, they defend each other from antibiotics, grow food together, and eat each other's waste. Microbes can send signals that determine the numbers of an adversary. In this way, the microbes find out if they are stronger than the defense system of the adversary. When a critical juncture is reached, microbes launch an attack. The attack can take many different forms.

Microbes are usually in constant communication with other microbes inside or outside their colony. They can intercept communication from other microbes and larger creatures. Microbes can change their behavior to defeat the larger creature or to help a comrade. Microbes can interpret the signals of other species and then send return signals to change the behavior of the other organism. Some microbes send signals to trick a rival into lowering its defenses. The signal can stop a yeast cell from growing into a more powerful multicellular fungus organism. In humans, some microbes increase the tendency of children to touch their mouths and thereby aid transmission of the microbe.

In a similar manner, human cells send signals to the bacteria and intercept their communications. A well-known example is the use of yogurt and other probiotics to send helpful signals in the human gut. Natural probiotics are colonies of microbes living in the gut that help the cells to maintain normal function and avoid infections. Incredibly complex communications occur all the time, as countless colonies of bacteria live in and around us. In fact, while a human has about 10 trillion of its own cells, it also has ten to a hundred times more bacterial cells and a thousand times more virus cells at any given time. The vast intercellular communication of microbes determines much of what happens in people.

MIT Professor Robert Rakes Schrock

MIT Professor M. Nafi Toksöz (The SEG Virtual Museum) writes:

> Trained in invertebrate paleontology, stratigraphy and sedimentology, Robert Rakes Schrock began his academic career at the University of Wisconsin where he remained from 1928 through 1937, the year in which he left Madison to join the geology faculty at MIT in Cambridge. It is Thomas Jefferson who wrote that 'no duty . . . is so trying as that of putting the right man in the right place.' The appointment of Schrock to MIT's Department of Geology was one of those extraordinary fortuitous occasions when the right man was placed in the right place and remained there for 38 fruitful years.

> During his chairmanship from 1949 to 1965, he brought about an extensive revision of the department's curriculum that involved strengthening the basic science requirements, adding subjects in geochemistry and geophysics, and establishing a

required summer field program. In 1956 he initiated together with Professor Houghton, head of meteorology, a joint program in oceanography with the Woods Hole Oceanographic Institution, a program which developed into the joint graduate-degree program that was formalized in 1968.

Soon after becoming head of the Department of Geology, Schrock, in tandem with Cecil Green (then president of Geophysical Service Inc.), organized a summer training program for geophysics students, the GSI Student Cooperative Plan, which operated for 17 years and gave some 350 students from 80 different US schools an opportunity to get practical field experience in geophysical exploration. The plan proved to be a most successful coupling of industry and academia and has remained a model for the cooperative educational effort. During all the years at MIT, Schrock was consistently outstanding in his ability to recognize and seize the right moments for change and expansion. His vision led to the establishment of the MIT Geophysical Analysis Group (GAG) project whose work has incontestably had a major impact on exploration geophysics.

Paleontology is the science of prehistoric life forms in the Earth's crust. It makes use of plant and animal fossils. Most people think of dinosaurs and other big prehistoric animals when they think of paleontology. However, the rock samples available to those engaged in drilling into the earth are in the form of "cuttings." Cuttings are small pieces of rock broken up by the drill bit. Cuttings contain an abundance of microfossils. By virtue of their small size, microfossils can be recovered whole. Microfossils tell us much about the life on earth over billions of years. There are a great number of different types of microfossils available for use.

Microfossils are underground but what about living things? One of the most significant discoveries is the presence of living microbes the layers of subterranean rocks. Such microbes are called endoliths. An endolith is an organism that lives inside rock, coral, animal shells, or in the pores between mineral grains of a rock. Many endoliths live in places previously thought inhospitable to life. Astro-biologists postulate that endolithic environments on Mars and other planets constitute potential refugia for extraterrestrial microbial groups.

On Earth, endoliths occur in rocks on the Earth's surface and in rocks miles within the subsurface. There are thousands of known species of endoliths, including members from bacteria, archaea, and fungi. Many endoliths are autotrophs, which mean that they are able to make their own organic compounds by utilizing gas or dissolved nutrients from water moving through fractured rock. Others may incorporate inorganic compounds found in their rock substrate, possibly by excreting acids to dissolve the rock.

Endoliths are found in a variety of environments, from the shallow surface to the deep terrestrial and ocean crust. Endoliths are a type of extremophile, which is an organism that thrives in harsh conditions. There are several different types of endoliths, each occupying a different environment. Recent work with endolithic geo-microbiology may alter our understanding of life. These microbes expand our understanding about the ability of organisms to survive and even thrive in extreme conditions. They also raise new ideas regarding the possibility of extraterrestrial life. Furthermore, these microbes may alter our ideas about the origin of life on Earth, because they have the ability to live within rock thereby escaping damaging UV rays.

Endoliths may also play a role in environmental issues. For instance, microbes living beneath the ocean floor have been suggested to play a role in the carbon cycle and global warming. Endoliths may also have environmental benefits including bioremediation of contaminated sites and mines and improvement of groundwater quality by converting harmful compounds into non-toxic waste products. In addition, these organisms may be responsible for bio-mineralization of economically important ores.

Remote sensing

Remote sensing refers to the acquisition of information about an object by the use of sensing devices that are not in physical contact with the object. Remote sensing is characterized by an intelligent use of signals that penetrate the unknown. There are two kinds of remote sensing: *passive* and *active*. In passive remote sensing, the observer waits for signals from unknown and unreachable regions. In active remote sensing, the observer emits signals into such regions and then records the resulting reflected, refracted, or scattered signals. Remote sensing depends on the use of signals that allow communication between the observer and the remote object. The most important type of signal is the traveling wave, whether mechanical (as in sound) or electromagnetic (as in light).

Many methods of remote sensing have been developed over the years. The development of the telescope and microscope in the early seventeenth century provided means to obtain images that could not be seen by the unaided eye. With telescopes, the astronomer can look outward to the stars and galaxies. With microscopes, the biologist can look inward to the microorganisms (microbes) that exist throughout the biosphere. An echocardiogram lets a doctor see the internal movements of the heart and blood without penetrating the skin. A satellite can monitor the features on the earth's surface and determine the environmental status of land cover, land use, and natural resources. Nondestructive testing finds hidden defects without taking the airplane apart.

If we could see the earth in cross section, we would find a sharp division between the core, or central part, and the mantle, or outer part. We would also find that the outer surface of the mantle has a very shallow skin layer of different composition, known as the crust. How did we obtain such knowledge? Earthquakes generate seismic waves. The seismic waves serve as the signals required for remote sensing. Because we must wait for earthquakes to occur, earthquake seismology represents passive remote sensing. In seismic exploration, source signals are transmitted into the ground, and the reflected seismic waves are recorded and processed. Exploration seismology represents active remote sensing.

Active remote sensing provides the basis for nondestructive testing. By comparing known input signals to measured output signals, the condition of a system can be determined without the use of invasive approaches such as disassembly or failure testing. For example, the system might be an aircraft wing whose trustworthiness must be routinely tested. Nondestructive testing does not require the disabling or sacrifice of the aircraft wing.

Noninvasive medical procedures, which involve no break in the skin, represent forms of remote sensing. Examples are magnetic resonance imaging (MRI), electrocardiography (EKG), electroencephalography (EEG), electrical impedance tomography (EIT), electroneuronography, magnetoencephalography, nuclear magnetic resonance spectroscopy, and diagnostic sonography (ultrasonography).

Remote sensing provides a way to search for the beginning of time. By the late 1920s, Edward Hubble had established that the universe was organized into galaxies of various sizes and shapes, each one consisting of billions of stars. It was found that most galaxies had light that shifted toward the red. The simplest interpretation of the red shift is that the galaxies are moving away from us. This conclusion fitted in well with Einstein's general theory of relativity.

Remote sensing makes it possible to collect data on dangerous or inaccessible areas. Remote sensing applications include monitoring areas such as the Amazon Basin and the Arctic and Antarctic regions, and the depth sounding of coastal and ocean depths. Remote sensing also replaces costly and slow data collection on the ground, ensuring in the process that areas or objects are not disturbed. Orbital platforms collect and transmit data from different parts of the electromagnetic spectrum, which in conjunction with large-scale aerial or ground-based sensing, provides researchers with enough information to monitor trends such as El Niño and other natural long and short term phenomena. Other uses include different areas of the earth sciences such as natural resource management and land usage and conservation.

A hologram creates a three-dimensional image of an object that we can see with our eyes. How can we create a three-dimensional of an object that we cannot see at all? Many ingenious methods of remote sensing have been developed for this purpose. These methods have opened up whole new worlds to us. The explorers of old traversed the seven seas to find new lands and then sailed home with their discoveries. We cannot travel to the stars to find out what they are made of. Instead, we must seek this information indirectly by analyzing starlight. We cannot travel underground to determine what is inside the earth. Instead, we must seek this information indirectly by sending seismic signals into the ground and interpreting their echoes.

Echolocation in animals

Several species of animals use echolocation for navigation and for foraging or hunting. The animal determines orientation to other objects through interpretation of reflected sound. Bats and dolphins are good examples of such animals. They emit calls and listen to the echoes of these calls. They use these echoes to locate, range, and identify objects. Echolocation works like active sonar. Ranging is done by measuring the time delay between the animal's own sound emission and any echoes that return from the environment. Unlike some sonar that relies on an extremely narrow beam to localize a target, animal echolocation relies on multiple receivers. Echolocating animals have two ears positioned slightly apart. The echoes returning to the two ears arrive at different times and at different loudness levels, depending on the position of the object generating the echoes. The time and loudness differences are used by the animals to perceive direction.

With echolocation, a bat can see not only where it is going but also how big another animal is, what kind of animal it is, and other features. Some moths have developed a protection against bats. They are able to hear the bat's ultrasounds and flee as soon as they notice these sounds, or stop beating their wings for a period of time to deprive the bat of the characteristic echo signature of moving wings which it may home in on. To counteract this, the bat may cease producing the ultrasound bursts as it nears its prey, and can thus avoid detection..

Echolocation is used by dolphins and most whales. It guides them through darkness of the deep and helps them to identify prey. Toothed whales emit a focused beam of high-frequency clicks in the direction that their head is pointing. Sounds are generated by passing air from the bony nares through the phonic lips. These sounds are reflected by the dense concave bone of the cranium and an air sac at its base. The focused beam is modulated by a large fatty organ known as the "melon." This acts like an acoustic lens because it is composed of lipids of differing densities. Most toothed whales use clicks in a series, or click train, for echolocation, while the sperm whale may produce clicks individually.

MIT Radiation Laboratory

Active remote sensing for distant objects requires waves to carry the information to and from the target. Radar is an active remote sensing system that uses electromagnetic waves to identify the range, altitude, direction, or speed of both moving and fixed objects such as aircraft, ships, motor vehicles, weather formations, and terrain. The term RADAR is an acronym for RAdio Detection And Ranging. The waves chosen for radar are microwaves. Such waves are electromagnetic waves in a frequency range above radio waves and below infrared waves. Microwaves provide the ranges and accuracies needed to detect aircraft and ships.

From September 1940 to May 1941 there were major aerial raids on sixteen British cities. British scientists had invented microwave radar to detect enemy aircraft. However, the magnetron in the British radar apparatus created microwaves with not enough power to reach the ranges desired. In September 1940, Great Britain sent a mission to the United States. The goal was to perfect and mass produce a magnetron tube to yield the microwaves with suitable power to do the job. Karl Compton, President of MIT, was the director of the section of National

Defense Research Council (NDRC) in charge of methods for the detection of aircraft and ships. Compton formed the MIT Radiation Laboratory.

MIT Professor Vannevar Bush was one of the founders of a small company named Raytheon. A meeting was arranged between Britain's leading scientists and Raytheon. Raytheon not only came up with radical changes that would simplify the manufacturing process, but also ways to improve the functioning of the radar. Impressed, Britain awarded, through the MIT Radiation Laboratory, "little" Raytheon a small contract to supply the magnetrons. At the same time it awarded "giant" Western Electric a large contract. By the end of the war, Raytheon was producing 80 percent of all magnetrons, leaving Western Electric, Radio Corporation of America (RCA), General Electric and other giant companies far behind. During the war Raytheon also developed the effective shipboard SG radar.

From 1940 to 1945, the MIT Radiation Laboratory made outstanding contributions to the development of microwave radar technology. Inventions included airborne bombing radars, shipboard search radars, harbor and coastal defense radars, gun-laying radars, ground-controlled approach radars for aircraft blind landing, interrogate-friend-or-foe beacon systems, and the long-range navigation (LORAN) system. Some of the most critical contributions of the Radiation Laboratory were the microwave early-warning (MEW) radars, which effectively nullified the V-1 threat to London, and air-to-surface vessel (ASV) radars. The MIT Radiation Laboratory closed on December 31, 1945. The MIT Lincoln Laboratory came into existence in 1952 in response to carry on the work of the Radiation Laboratory into new directions. Lincoln Laboratory developed radar systems to make travel by airplane safe.

After the war, Raytheon, needing something to do, came up with microwave cooking. In 1947 Raytheon demonstrated the world's first microwave oven. For the Office of Naval Research, Raytheon built the Hurricane computer. In 1948, Raytheon became the first company to develop a missile guidance system that could hit a flying target. In 1948,

Raytheon also released the first commercially produced transistor, the CK703 point contact transistor. Raytheon fo lowed this by the CK722 germanium junction transistor, the first transistor sold to the public. In 1954, Texas Instruments released the first ccmmercially produced transistor radio. Raytheon responded the next year with its own version of the transistor radio, the 8TP-4. Transistors were to completely revolutionize computers and electronic technology.

Reflection seismology

Sound will reflect off a wall. Stand away from a large flat wall and clap your hands repeatedly. Almost immediately you will hear the echo (i.e., reflection) of your clapping, slightly out of step with it. The echo is the sound energy in your clap traveling out to the wall, bouncing back, and eventually entering your ears. The delay between the sound and the echo represents the time for the sound to go to the wall and back. The larger is the distance, the longer is the delay.

Refection seismology operates on the echo principle. In the early days dynamite was used as the source of energy. A shot of dynamite emits a source pulse which travels down into the earth where it is reflected from the subterranean interfaces between rock layers. Receivers (a.k.a. seismometers or geophones) on the surface ɔick up the reflected pulses. In the early 1950s the received signals were recorded as traces on a strip of moving photographic paper. (By the late 1950s the received signals were recorded as analog signals on magnetic tape). Each trace on the paper represented the response of one receiver to the shot. In addition to the energy derived from the source, the seismometers also pick up energy not derived from the source. This non-source energy would include such things as wind noise and highway traffic noise. However, such extraneous energy is generally held to a low level. It is usually so minor that it is not significant on seismic traces. As a result, virtually all the energy on an exploration record is due to the shot.

If you look straight at a mirror, you see yourself. Your image is a primary reflection. If you look straight at a second mirror, you see another primary reflection of yourself. If the second image is smaller than the first image, then you can infer that the distance to the second mirror is greater the distance to the first mirror. For this reason primary reflections are desirable for the purposes of remote sensing.

The Palace of Versailles is known for its luxurious furnishings, intricate gardens, and overall extravagance. Its Hall of Mirrors, completed in 1686, features breathtaking garden views through seventeen ornate windows. The room has 578 mirrors, which are of exceptional size. Mirrors hang on the walls opposite the windows, strategically placed to reflect the natural light. A series of sparkling chandeliers adorn the ceiling. The ceiling is decorated with paintings celebrating the first years of the reign of King Louis XIV. The Hall of Mirrors represents the society of the royal court, in which seeing and being seen were crucial. In the Hall of Mirrors, every movement, every nod, every glance was reflected hundreds of times. These multiple reflections jumbled the images that were seen. The dazzle was amazing, but the stakes were high: a stumble, a glance, an awkward step, could be magnified in ways that would be misleading, even dangerous. For the same reason multiple reflections were misleading, even dangerous on the paper seismic records of the 1950s.

Let us explain. A reverberation is a special kind of multiple reflection. The category "multiple reflections" includes both reverberations and other types of multiple reflections. A reverberation in a multiple reflection that makes many repetitive turns between the same two interfaces. For example, in exploration at sea, the water layer acts like a drumhead. A drumhead is a membrane stretched over one or both of the open ends of a drum. The drumhead is struck with sticks, mallets, or hands, so that it vibrates and the sound reverberates through the drum. Because water reverberations generally overwhelm the primary reflections from depth, exploration at sea was not possible (except in rare instances) in the days before the digital computer.

In music we like reverberations. Composers of sacred music made use of the complex natural reverberations inside cathedrals. This knowledge was later used in the design of opera houses and concert halls. They were built to create reverberations that would enhance sound in the days before electrical amplification.

In seismic analysis, we do not like reverberations because they hide the primary reflections. We need to remove the reverberations (and other multiple reflections) in order to detect the primary reflections. The primary reflections represent the signal. Remember that we have assumed that the ambient noise is so small that we can rule it out. As a result the reverberations (and other multiple reflections) represent the noise. If there were no primary reflections, there would be no multiple reflections. In other words, if there were no signal, there would be no noise. That is why reverberations (and other multiple reflections) are called signal-generated noise. We need to deconvolve the received seismic trace to take away the signal-generated noise so that we obtain the signal.

Most of the underground structure of the earth's near crust is like a vast and elaborate cathedral, even more multifarious than Chartres. In such complex situations, the reverberations would so overwhelm and hide the primary reflections that analog seismic method would not work. In 1950, such regions of the earth were not explorable. In a relatively few areas the underground structure was like a plain unadorned colonial New England meeting house. In such situations, the analog seismic method did work but at the cost many dry holes.

ADDED NOTE: From the original seismic computations on Whirlwind in 1952, the seismic exploration industry has expanded its capabilities. A recent news item states: BP's computing requirements are now 40,000 times greater than they were in 1999. Every second, BP's seismic supercomputer in Houston does nearly 4,000 trillion floating-point operations. That is 3.8 petaflops of computing speed. Just a few years ago, BP's now-retired supercomputer was the first commercial machine to reach one petaflop. This facility opened in 2013 with double that speed, and has doubled again since—a nearly exponential acceleration. The plan is to continue that pace of growth by continually upgrading the leased servers as newer ones become available, and adding more.

Chapter 4. Dead reckoning

> Max Planck: We are in a position similar to that of a
> mountaineer who is wandering over uncharted spaces, and
> never knows whether behind the peak which he sees in front of
> him and which he tries to scale there may not be another peak
> still beyond and higher up.

History of oil exploration

The history of oil exploration starts out with the history of drilling. An oil well is a hole that is dug to bring oil to the surface. Usually some natural gas is produced along with the oil. A well that is designed to produce mainly gas is called a gas well. In the early days oil fields were discovered by drilling at oil and gas seeps. Geological hunches and petroleum residue in soil samples were responsible for discoveries. In the 1920s, geophysical methods began demonstrating an advantage for finding oil. The geophysical tools made use fundamental variables such as gravity change, magnetic field change, time change, and electrical resistance. The two principle seismic methods are refraction and reflection. The seismic method was developed in the 1920s and came into general use by about 1930. The Geophysical Research Corporation began experimenting with the seismic reflection method in 1926 and by 1929 had seismic crews employing the method commercially throughout West Texas and the Gulf Coast. Oil exploration is an expensive, high-risk operation. Generally the most costly aspect is getting possession of favorable land. The drilling of wells is costly. The geophysical exploration of the land is the least costly but by far the most critical aspect. All is lost if the geophysics fails. And fail it did. The only recourse was to keep drilling more wells. By 1950 great efforts were being made to improve the seismic method.

In 1950, good records resulted in those instances when the signal-generated noise (reverberations and multiple reflections) was low. In such cases the interpreters could see the signal (primary reflections) on the traces. Poor records resulted when the signal-generated noise was high. In such cases the interpreters could not see the signal (primary reflections) on the traces. The problem became one of uncovering (by eye) the primary reflections hidden in the confusion of the signal-generated noise. In other words, the problem was one of visually separating signal (primary reflections) from signal-generated noise (reverberations and multiple reflections). Unfortunately, most geologically promising areas of exploration yielded seismic records on which few if any reflections could be picked visually. The answer before was to just keep drilling and drilling. The United States Steel Company was making a fortune on the sale of steel drilling pipe. Even more spectacular than the sight of many oil derricks was the sight of huge storage areas with tons and tons of steel pipes ready to be used.

The *Rime of the Ancient Mariner* by Samuel Taylor Coleridge, written in 1797-1798, relates:

> How a Ship having passed the Line was driven by storms to the cold Country towards the South Pole; and how from thence she made her course to the tropical Latitude of the Great Pacific Ocean; and of the strange things that befell.

The version of 1817 features a gloss. A gloss is a brief marginal notation of the meaning of the text. Here we take a few stanzas from the poem, but insert our own gloss in brackets.

[In rounding Cape Horn with "ice mast-high floating by," the ancient mariner shoots the Albatross. As the symbol of this recklessness, the dead Albatross is hung about the neck of the ancient mariner. By way of analogy, the nations of the world were "parched" for oil in **1950**.]

> There passed a weary time. Each throat
> Was parched and glazed each eye.
> A weary time! A weary time!
> How glazed each weary eye,
> When looking westward, I beheld
> A something in the sky.

[All of a sudden, the digital computer (Whirlwind) appears in **1951**.]

At first it seemed a little speck,
And then it seemed a mist;
It moved and moved, and took at last
A certain shape, I wist.

A speck, a mist, a shape, I wist!
And still it neared and neared:
As if it dodged a water sprite,
It plunged and tacked and veered.

[In **1953** geophysicists started "drinking all" of what digital could offer.]

With throats unslaked, with black lips baked,
Agape they heard me call:
Gramercy! they for joy did grin,
And all at once their breath drew in,
As they were drinking all

[However by **1954** cyberanxiety in geophysics took hold. The digital computer was a "spectre bark," an unreliable omen.]

The sun's rim dips; the stars rush out:
At one stride comes the dark;
With far-heard whisper, o'er the sea,
Off shot the spectre bark

[In the **1960s**, digital processing started it ascendency. Great new oil fields were discovered. However, petroleum now began to be feared.]

Beyond the shadow of the ship,
I watched the water-snakes:
They moved in tracks of shining white,
And when they reared, the elfish light
Fell off in hoary flakes.

[By drilling into the crust of the earth, people became knowledgeable of the long and intricate history of the earth. Petroleum (i.e., naturally occurring hydrocarbons formed from living things throughout geologic time) is an integral part of the life cycle that makes our planet special. It represents the abundance of living things, their beauty and wonder. As a battery stores electric energy, petroleum stores the energy of animal life. This stored animal energy could rejuvenate bacterial life on earth in case of catastrophe, as from volcano activity or asteroid encounter. The computer (artificial life) is used to find petroleum (fossil life)].

O happy living things! no tongue
Their beauty might declare:
A spring of love gushed from my heart,
And I blessed them unaware:
Sure my kind saint took pity on me,
And I blessed them unaware.

The self-same moment I could pray;
And from my neck so free
The Albatross fell off, and sank
Like lead into the sea.

The skill of the steersman

In navigation, dead reckoning is the process of calculating the current position by using a previously determined position, and advancing that position based upon known or estimated speeds over elapsed time and course. One of the most pronounced eras of geographic exploration was in the Age of Discovery which started in the time of Christopher Columbus. Explorers sailed and charted much of the world. Space exploration started in the 20th century with the perfection of artificial satellites and other space vehicles. Also explorations to the depths of the oceans started in earnest in the twentieth century. With robotic machines, humans have explored parts of the heliosphere, and through measurements, beyond the Solar System and beyond the Milky Way as part of an ongoing global Space exploration initiative.

Fortunately the logbook of the first voyage of Columbus was preserved. Few other logbooks of those early days of exploration are extant. Columbus' logbook shows that he navigated by dead reckoning. At sea Columbus would be out of sight of land so he would have no reference to any visual landmark. Starting from a known fixed point (the Canary Islands), the great navigator measured the course and distance of each of the successive legs of the journey. Direction was measured by a magnetic compass. Distance was determined by a time and speed calculation. The ship's page faithfully turned the hourglass to keep the time. The navigator multiplied the estimated speed of the vessel by the time traveled to get the distance. The officer of the watch would keep track of the speed and course sailed every hour. At the end of the day, the total distance and course for the day was transferred to the chart. The surviving logbook of Columbus shows the measurements of course and distance sailed. If Columbus had been a celestial navigator, we would expect to see records of celestial observations. However, Columbus's logbook does not show celestial records. Columbus relied upon dead reckoning.

In 1848, the school census showed that San Francisco had a population of 575 males, 177 females and 60 children, a total of 812. In May of that year Sam Brannan, a storekeeper in Sutter's Creek, brandished a bottle filled with gold dust on the streets of San Francisco shouting "Gold! Gold! Gold!" It was the spark that ignited the California gold rush. The stampede was on. Jeremy B. Wardwell, whose family originally came from Alford, Lincolnshire, heard the cry when he was 18 years old. He sailed from Boston, Massachusetts, on November 28, 1849 on the square-rigged ship *Nestor*. The beautiful wave-parting *Nestor* was built in Portsmouth, N.H. The ship carried a large load of lumber and machinery in addition to gold-seeking passengers. The development of the marine chronometer (a mechanical clock) in the 18th-century meant that ships could be much more accurately navigated than in the days of the hourglass. Unfortunately, part of the Wardwell's manuscript is missing, and it commences abruptly in the midst of a great storm in rounding Cape Horn. The following passage illustrates that it was the skill of the steersmen that saved the ship.

A terrific Northwester burst upon us. The sea broke over us filling our decks, poured over the main hatch combings, and rushing down between decks. As the ship rolled heavily, many inches of water had accumulated on the lower deck rushing in quite a wave, carrying boxes, kegs &c &c across the deck from side to side. The temperature fell low and the spray froze to the rigging and made the deck slippery. It was a terrific night for the sailors, some of whom were green hands (balance of page faded out — account continues) . . . reef main topsail and fore topmast staysail, for fifty hours. When running on one of those tremendous wave mountains with the ship head depressed at a sharp angle, it seemed as though she must run under and be overwhelmed instead of ever rising again. Captain Nathan Pool was a good and careful sailor, and often during that gale I saw him spring to the wheel and add his whole strength to that of the two men who were steering to prevent the ship from broaching sufficiently to get broad side to the trough of the sea in which case she would have lost her headway and became unmanageable, then nothing could have saved her from destruction. No serious damage was done during this terrific gale, excepting the bursting of the main topsail, which was repaired aloft while the ship was scudding. It was indeed a trying situation for those sailors who were engaged in repairing that sail. From the breaking crests of those enormous waves, the wind carried the spray into the air until it was filled white with it, as light snow fills the air during a gale upon the land. For a time all went well, and we finally reached the region of calms.

Cape Horn is notorious as the most dangerous ship passage in the world. Many ships were wrecked attempting to round the Horn. The waters are particularly hazardous, owing to strong winds, large waves, strong currents and icebergs. The Westerlies are the prevailing winds in latitudes below 40° south. These hazardous winds blow from west to east around the globe almost uninterrupted by land. They create particular problems for vessels trying to round the Horn against them, i.e. from east to west. Northwesters (terrible storms that occur at Cape Horn) would drive square-rigged ships south into regions of icebergs and sea ice, a graveyard for ships. In a rounding that was described as horrific, the ship *Susanna* beat against the wind for 91 days, falling at one point as low as 62 degrees south. The *Nestor* rounded Cape Horn falling to latitude 61 degrees south. Latitude 61 degrees south is the same as that of Elephant Island, an ice-covered mountainous island situated 152 miles north-northeast of the tip of the Antarctic Peninsula. The island is famous as the desolate refuge of the intrepid British explorer Ernest Shackleton and his crew in 1916.

Wardwell engaged in placer mining on the Tuolumne River near Hawkins Bar during the ill-fated summer of 1850. Like hundreds of his fellow miners Wardwell saw his summer's work in dam construction washed downstream on the crest of a sudden freshet. A passage taken from Wardwell's manuscript is as follows.

> We arrived at San Francisco all well in one hundred and eighty-eight days from Boston. At that time San Francisco was mainly a city of tents. It had previously suffered much from fires. Merchants were doing a heavy business in rough board buildings of one storey. I had no occasion to purchase lumber but was informed that it was worth $200 per thousand. How often in my imagination have I made that voyage in the "Nestor." Sailing down the Southern slope of the Atlantic Ocean till passing Cape Horn, then on a Westerly course gradually rising higher and higher on the grand Pacific till we reach a Northern latitude nearly as high as that from which we sailed six months before. We are however 51 degrees West of our point of departure. With light hearts and buoyant hopes, we danced on, plunging the mountain waves which many times threatened destruction, and casting a gloom over the ship's company.

Wardwell came back to Boston after five years in the gold fields. Most accounts of the gold rush only deal with people and their struggles. In the words of an historian, Wardwell's account of California river mining serves to "add much to the sparse knowledge of how this activity was carried on in 1850."

Inertial navigation (INS) vs. global positioning (GPS)

MIT Professor Charles Stark Draper (1901-1987) is known as the "father of inertial navigation". He was the founder and director of the MIT Instrumentation Laboratory, later renamed the Charles Stark Draper Laboratory. On 25 May 1961 President John F. Kennedy announced the Apollo program to send humans to the Moon. The MIT Instrumentation Laboratory received the first contract for the program. The Laboratory designed the Apollo Guidance Computer. It was a one-cubic-foot computer that controlled the navigation and guidance of the Lunar Excursion Module to the Moon on nine launches, six of which landed on the Moon's surface.

Draper invented and developed inertial navigation. An inertial navigation system INS (a.k.a. inertial guidance system) is a device that does dead reckoning for a moving vehicle. The device has a computer, motion sensors (accelerometers) and rotation sensors (gyroscopes). Its computer continuously calculates the position, orientation, and velocity (direction and speed of movement). An inertial guidance system does not need any external reference. It only uses measurements provided by its self-contained accelerometers and gyroscopes. The device tracks the position and orientation of the vehicle relative to a known starting point, orientation and velocity. Because the device makes use of no external signal, it cannot be jammed. It is used on vehicles such as ships, aircraft, submarines, guided missiles, and spacecraft. Other terms used to refer to inertial navigation systems include inertial guidance system, inertial instrument, inertial measurement units (IMU). Recent advances in micro-electromechanical components make possible small and light inertial guidance systems.

The chief characteristic of the inertial navigation system (INS) is that no external signal is needed. In contrast, the chief characteristic of global positioning system (GPS) is that external signals are needed. The GPS is a satellite-based navigation system made up of a network of 24 satellites placed into orbit by the U.S. Department of Defense. The GPS satellites circumnavigate the earth twice a day and transmitting signals to earth. The GPS device uses this information to calculate the user's location. Newer GPS devices have an average accuracy of three meters. The 24 GPS satellites are powered by solar energy with batteries for backup. Rocket boosters on each satellite keep them in the correct path. A happy combination is GPS/INS. It uses of both global positioning and inertial navigation. GPS satellite signals are used to correct or calibrate the position given by an inertial navigation system (INS). Inertial navigation systems are usually subject to drift, so they can provide accurate positions only for a short period of time. The GPS gives an absolute drift-free position value that can be used to reset the INS solution.

The brain of many animals has an inherent navigational system. Salmon spend their early life in a river, and then swim out to sea for their adult life. When mature, they return to the same river, and even to the same spawning ground. From the Sargasso Sea, young American eels drift with ocean currents and then migrate inland into streams and lakes. They feed for approximately 10 to 25 years before migrating back to the Sargasso Sea in order to spawn. Homing pigeons, after being transported, can find their way back from distant places that they have never visited before. Sea birds, such as the albatross, can fly home from as much as 4,000 miles away. Monarch butterflies annually migrate 2,000 miles from the Great Lakes to Mexico and back. No individual butterfly completes the entire round trip. Female monarchs lay eggs for the next generation during the northward migration and at least five generations are involved in the annual cycle. However the Monarch butterflies return every winter to the same "butterfly trees" in Mexico. A cat or dog transported from its home to an unfamiliar location, with no chance to learn smells or landmarks en route, can find its way home. An ant or bee does not return along the same tortuous outgoing path by which it has come. An ant or bee computes it location continuously so it can always make a beeline (i.e., a straight direct course) for home.

Lowly but highly resourceful ant

"Allo" is a combining form indicating difference, variation, or opposition. The word *allothetic* refers to information concerning an animal's orientation in an environment that s obtained by the animal from external spatial clues. "Idio" a combining form meaning one's own, private, personal. The word *idiothetic* refers to information concerning an animal's orientation in an environment that is obtained by the animal by reference to its previous orientations and movements, and without external spatial clues.

The explorers of old used dead reckoning. What is dead reckoning? In navigation, the steersman sees allothetic clues, such as land and other ships. The steersman uses these allothetic clues as feedback to continuously correct his course. Suppose that the ship is alone at sea out of sight of land. Now the steersman can use no feedback. He has no choice but to use idiothetic clues, such as his starting point and his measurements of direction, speed, and time. The steersman must use these idiothetic clues as feedforward to continuously correct his course. In summary, **allothetic clues** are external clues that can be used as **feedback** to determine the correct action. **Idiothetic clues** are internal clues that can be used as **feedforward** to determine the correct action.

Dead reckoning in animals is usually known as path integration. Animals, of course, do not have the instruments that Columbus used for dead reckoning. Often the animal cannot see any landmarks, such as on dark nights, or on the open ocean, or in featureless areas such as deserts. In such case, the animal must rely on idiothetic clues from within its body in order to do path integration. The idiothetic clues come from different sensory sources within its body. The animal must be able to put together these internal clues, without making reference to visual or other external landmarks. The animal estimates position relative to a known starting point (e.g., its nest) continuously while travelling on a path that is not necessarily straight. Path integration is used by ants, bees, birds and rodents. At any point along its journey, such an animal can always return directly in a straight line to its nest.

One idiothetic cue in animals comes from the vestibular organs, which detect movement in the three dimensions. Other idiothetic clues include proprioception (information from muscles and joints about limb position), motor efference (information from the motor system telling the rest of the brain what movements were commanded and executed), and optic flow (information from the visual system signaling how fast the visual world is moving past the eyes). Together, these sources of information can tell the animal which direction it is moving, at what speed, and for how long. In addition, sensitivity to the earth's magnetic field for underground animals (e.g., mole rat) is used in path integration.

Sahara desert ants forage for insects that have died of heat stress. These ants can sustain surface temperatures of up to 70 °C (178°F). This ant travels far from its nest. The ant will venture out a distance as much as a third of a mile on a tortuous path that continual meanders and twists in all directions. The Sahara desert sand has almost no identifiable features. The ant completes its journey when it finds a dead insect. By path integration (feedforward), the ant goes directly back to its nest by the shortest route. This skill is necessary to its survival under the harsh desert conditions. With respect to size, this ant travels farther from its nest than any other animal that lives on the Sahara. An ordinary GPS receiver would have an accuracy of about 12 feet, not enough for an ant to find its nest.

There are no landmarks on a desert. It is all sand. An ant does not know where or when it will find a dead insect. As a result, at each point along its path, the ant computes the distance and direction to its nest. How does the ant do this? Ants can do this in the dark, so they do not need the sunlight. It seems that the ant counts the steps it takes. At the same time the ant must correct for terrain. It takes more steps to traverse a fixed horizontal distance on a bumpy terrain than on a flat terrain. However the ant gets the correct value in either case. In summary an ant mentally can make a hilly terrain level and, on the way back, make a circuitous path straight. It fulfills *Luke 3:5:* "Every valley shall be filled, and every mountain and hill shall be brought low; and the crooked shall be made straight, and the rough ways shall be made smooth."

Prior and during World War 2, MIT built and continually improved the 100-ton analog computer known as the differential analyzer. It was the largest and most intricate analog computer ever made. It was used to determine the paths of artillery shells. It did so by path integration implemented by means of wheel-and-disc integrators. In comparison to the ant's tiny brain, the differential analyzer could not hold a candle. The ant indicates that it may be possible to build extremely small and powerful analog computers in the future.

Path integration is a feedforward process because it does not make use of any external landmarks. Effective feedforward devices are extremely difficult to construct. Experiments have been conducted with desert ants. At the dead insect, the ant is put to sleep. Its legs are either cut short or else elongated by stilts that are glued on. The dismembered ant is displaced so it will not walk over the nest on its return journey. When awakened an ant with stumps walked too short and an ant with stilts walked too far on the way back. The dismembered ant uses the nest as an external landmark to make a feedback correction. On the very next journey the ant with either stumps or stilts goes out to find a dead insect and walks exactly the right distance on the way back.

The desert ant represents an extreme case. The ant is alone on the sand. The ant walks about looking for a dead insect in the overwhelming heat. It navigates not by external landmarks but by internal sensors. Once the ant finds a dead insect, or once the ant gets too hot, the ant must make a beeline back to its underground nest. The word "beeline" means "to go quickly in a direct course." It is an apt description because bees are also governed by the same type of navigation system as the ant. The ant keeps track of its entire course history from the time it leaves the nest. The ant's brain feeds forward all of this history to determine the current beeline back to the nest. This determination does not make use of any knowledge provided by the surroundings. The navigation system of the ant is a purely feedforward process. A person walks about Manhattan all day long. He does not keep track of his course in any way. When he wants to go home he looks at the street sign. He calls home and says, "Pick me up at 61th street and Broadway."

This determination only makes use of knowledge provided by the surroundings. The navigation system is a purely feedback process.

Biological clocks

Biological clocks are natural devices existing in living things that keep track of time. Every organ and even every cell seems to have some sort of clock. The clocks preserve the daily patterns of activity when the organism is removed from daily light or temperature changes. It is not known how biological clocks work. Does the organism have a physicochemical timing system that generates periods of the same duration as those of the earth and moon? When deprived of the usual environmental variations such as those in light and temperature, can the organism detect related variations in such forces as the earth's magnetic and electric fields? Biological clocks in humans keep the daily rhythm of sleep and wakefulness. Such rhythms are called circadian (i.e., about a day). Circadian rhythms can vary widely among living things. Underlying this rhythm there are daily variations in blood sugar, kidney excretion, cell-divisions, body temperature, and hormone secretion. When humans are confined in rooms without clocks or windows, they try to keep daily patterns of sleep and wake. However they systematically drift to progressively earlier or later times of day. The resulting rhythmic periods will deviate somewhat from 24 hours.

Biological clocks occur throughout the animal kingdom, from single-celled forms to mammals. Animal clocks adjust activities to the day-night cycles or to the ebb and flow of the tides. They also time monthly and annual migratory and reproductive activity. The annual return of the swallows to Capistrano, California, occurs on the same day each. The annual breeding of the palolo worms of the South Pacific follows a timed rhythm. These worms rise to the surface of the water for breeding each year at dawn exactly one week after the November full moon. Plants also possess clocks. Bean seedlings raise their leaves in the morning and lower them at night. The brown alga *Dictyota* has a monthly reproductive rhythm. Daily variations in respiration, photosynthetic capacity, and growth rate are governed by internal biological clocks.

Chapter 5. Analog computers

Max Planck: Science cannot solve the ultimate mystery of nature. And that is because, in the last analysis, we ourselves are part of nature and therefore part of the mystery that we are trying to solve.

Analog computer and digital computer

Instinct refers to an innate, typically fixed pattern of behavior in animals in response to certain stimuli. In contrast, learning is the acquisition of knowledge or skills through experience, study, or being taught. The ability to simplify complex concepts with analogies is a critical skill. This skill is called learning by analogy. At some point a child might witness an animal, like a bird or a squirrel, die. The child reasons, "I could die as well." It is well known that children learn complex concepts best by analogy and by simplification. In fact, analogies are essential in the understanding of complex scientific concepts. In physics students are taught that an electrical network is like a network of pressurized water pipes. After such an analogy is made, the teacher must then go on and describe the essential difference between the two concepts that are compared. Reasoning by analogy generally involves abstracting details from a particular set of problems and resolving structural similarities between otherwise distinct problems.

An analog computer computes one thing by making an analogy to another. It does not count in discrete units. It uses the continuously changeable aspects of physical phenomena such as electrical, mechanical, or hydraulic quantities to solve a mathematical problem. An analog computer mimics mathematical equations by voltages or gear movements. It then performs calculations by comparing, adding, or subtracting voltages or gear motions in various ways. The final result is sent to an output device such as a screen or printer. Analog computers can have a very wide range of complexity. Slide rules were the simplest and most commonly used, while military gunfire control computers were among the most complicated.

A digital computer represents varying quantities as discrete numbers. It performs calculations based solely upon these numbers. A digital computer requires an analog-to-digital converter (abbreviated ADC) to turn analog signals into digital form for processing. The computer processes the digital data. By the use of a digital-to-analog converter (abbreviated DAC) the processed signals can be turned back into analog form. In the 1950s and 1960s, digital computers came on the scene. Their success made analog computers largely obsolete. However analog computers remain in use in specific applications. However natural analog computers found living things exceed the capabilities of any digital computer.

Analog computers in nature

Bees routinely solve complex mathematical problems that would keep a digital computer busy. In particular, the bees learn to fly the shortest route between flowers discovered in random order, effectively solving the "travelling salesman problem." The problem involves finding the shortest route that allows a travelling salesman to call at all the locations to be visited. Digital computers solve the problem by comparing the length of all possible routes and choosing the one that is shortest. Would the bees follow a simple route defined by the order in which they found the flowers, or would they look for the shortest route? After exploring the location of the flowers, the bees quickly learn to fly the best route for saving time and energy.

Despite their tiny brains bees are capable of extraordinary feats of behavior. Bees manage to reach the solution using a brain the size of a grass seed. Foraging bees solve traveling-salesman problems every day. They visit flowers at multiple locations and, because bees use lots of energy to fly, they find a route which keeps flying to a minimum. We have no idea how a bee can solve the travelling salesman problem without a sizable digital computer. There is no digital computer yet made that can solve the problem as fast as a bee can. The extraordinary brain of a bee is essentially an analog computer. Here is a case where an analog computer easily outstrips a digital computer. Prof. Wiener's vision that we must study both animal and machine represents the essence of cybernetics.

Analog computers in engineering

A computer is device designed to accept data, perform prescribed mathematical and logical operations, and display the results of these operations. An analog computer is a computer that works with analog quantities. An analog clock is a device in which the hands go around in a circle. It is an analogy of the apparent motion of the sun going around the earth. Although there are twelve numbers on the face of the clock, we do not regard the clock as a digital computer. Of course the mathematical equations of Kepler and Newton explain the movements of both the sun and the clock. However, the people who designed the clock came before Kepler and Newton and so could not make use of their equations.

Analog computers essentially deal with mathematical and logical operations. The analog computer solves equations, not knowing what the equations represent. The same equations might represent either the flow of water or the flow of traffic. It makes no difference to the computer, which just blindly goes about solving the equations. The physical phenomenon that produced the equations is not taken into account. The analog computer follows the equations, not the physical phenomenon. A human must analyze the physical situation and then formulate the equations for the computer. The most advanced analog computer ever made was the differential analyzer at MIT. It was designed especially to solve differential equations. Newton's law of motion, Maxwell's electromagnetic equations, and Einstein's equations of general relativity are all differential equations. In fact, nearly all the equations of mathematical physics are differential equations or can be reduced to differential equations.

The earliest hand-held analog computer is the astrolabe. First built in Greece around the second century BC, the astrolabe uses gears and scales to predict the motions of the Sun, planets, and stars. The slide rule is a hand-held analog computer that was used up to about 1970, when the digital hand-held calculator took over. A ruler is a straightedge with calibrated lines measuring distances. One of the simplest analog computers would consist of two 12 inch rulers each with linear markings from 0 to 12. Place both rulers next to each other. Suppose we want to add 3 and 4. Put your left finger at 3 on the lower ruler and slide the 0 on the upper ruler to your left finger. Then put your right finger on the 4 on the upper ruler and look down. You will see 7 on the lower ruler, which is the answer.

In the sixteenth century, scientific progress was at a standstill largely because of the difficulty and toil of numerical calculations. The situation changed when John Napier (1550–1617), born in Edinburgh, discovered logarithms. Tables of logarithms were essential in science and engineering until the introduction of hand-led calculator in about 1970. The slide rule was invented shortly after Napier's discovery. The magic rule is: The logarithm of the product of two numbers is the sum of the logarithms of the individual numbers. The slide rule consists of two rulers as described above. However, the scale now is not in inches but in the logarithms of inches. The slide rule does multiplication by adding logarithms. It does division by subtracting logarithms. In effect, the slide rule is a hand-operated analog computer.

Analog computers are concerned with computation. Physical quantities such as distance between marks on a ruler, electric currents through wires, angles through which a disk is rotated, or light transmitted through optical material can be used to represent numbers. By exploiting known physical properties, mathematical operations are then performed to give the desired results.

Any numerical computation basically a series of additions, subtractions, multiplications, and divisions performed according to a predetermined plan. Since these four arithmetical operations can be performed by mechanical, electrical, optical, and other means, it is possible in theory to make analog computers capable of performing any computation. However, it can be difficult, and often extremely difficult, to implement the theory. The construction of analog computers is the art and practice of assembling physical components. Addition and subtraction can be constructed is relatively easy ways. For example, we may add or subtract two numbers by representing them as currents in wires and then merging them in parallel or antiparallel directions. It is more difficult to implement multiplication and division. We can multiply two electric currents by feeding them into the two magnets of a dynamometer. The result is a rotation, which can be transformed back into current which is the product of two input currents. Unfortunately such a mechanism is complicated to build.

One of the great advances was the invention of the differential gear, which acts as an analog computer. It is used when the back axle of an automobile provides the power from the engine to the wheels. When the car turns, the wheels on the outside have to spin faster than the ones on the inside because they have to travel a longer distance. If you had a fixed axle (where one wheel cannot spin independently of the other) one of the wheels will have to skid if the axle is going to keep turning. The differential mechanism basically allows each wheel to spin freely of the other, while providing power to both.

The mileometer (a.k.a. odometer) is a device that records the number of miles that a bicycle or motor vehicle has travelled. Let us explain. As the automobile wheel rolls along any curved road, the milometer counts the number of turns the automobile wheel makes. By multiplying the number of turns by a conversion factor, the distance that the automobile traveled is obtained. In a similar way a wheel turning on a surface can be used to achieve the mathematical operation know as integration. More specifically, let the wheel turn on disk that itself rotates. This mechanism gives what is called a wheel-and-disk (or ball-and-disk) integrator. It can be made to compute any kind of mathematical integration. It is the basic component for a mechanical analog computer.

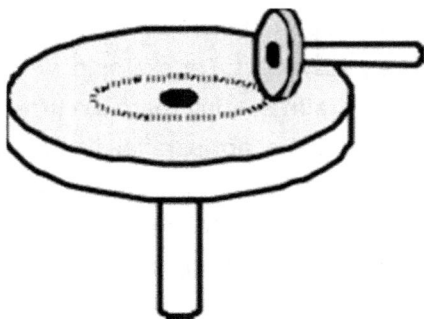

Figure 2. Wheel-and-disk integrator.

The wheel-and-disk integrator (Figure 2) is a variable-speed gear. A small knife-edged wheel rests on a rotating horizontal disc. The vertical axis of the horizontal disc is supported in a movable carriage so that the distance of the point of contact of the wheel on the disc, from the center of the disc, can be varied. The two inputs to the integrator are therefore the disc rotation and the carriage movement. When the unit is integrating, the disc rotates and undergoes a translational movement simultaneously. The rotating disc drives the wheel by friction. The rotation of the wheel represents the output of the integrator. Integrating with respect to another variable is the nearly exclusive province of mechanical analog integrators (the wheel-and disk mechanism); it is almost never done in electronic analog computers.

Slide rule as analog computer

The computer that built America, the slide rule, is now extinct. The slide rule was the most successful analog computer ever made. In the 1940s every student at MIT had one. Until the advent of modern navigational technology, air navigation in the dark was intensely difficult, particularly if there were cloud-cover over the ground. Crews on airplanes had to rely on dead reckoning: estimating position by speed, flying time and compass. The dead reckoning was done by using a slide rule. As in the voyages of Columbus and the other great steersmen in the age of exploration, unpredictable winds and many other things could disrupt the finest calculations.

In the 1940s on an airplane, unlike on a ship, the navigator had to make many calculations and had to make them in a hurry. Computers (i.e., slide rules) were used to solve the problem accurately, and it could be done quickly and easily. The E–6B computer was designed to be used for nearly all computations in air navigation.

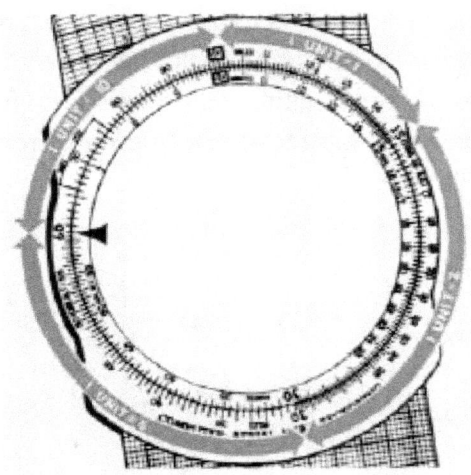

Figure 3. Slide rule on back side of the E–6B computer

The circular slide rule (Figure 3) on the back of the E–6B computer has a stationary outer scale, and an identical movable inner scale on a rotating disc. In calculations, the outer scale usually represents units of measure (miles, gallons) while the inner scale represents units of time (hours and minutes). However since these scales are standard logarithmic scales they can be used to solve any problems in multiplication or division.

Distance, Time and Speed. The black pointer of the inner (time) scale is used in all problems involving time. There is a definite relationship between distance, time, and speed. Distance is the product when time is multiplied by speed. Speed is the quotient when distance is divided by time. Time is the quotient of distance divided by speed.

Fuel consumption. Fuel consumption problems are calculated in the same manner as distance, time, and speed. The only difference in the problem is that you substitute gallons of fuel for units of distance.

True altitude. When you use the E-6B to compute your true altitude for navigation, always be sure to use the window marked FOR ALTITUDE CORRECTION. Adjust rotating disk to bring flight level pressure altitude opposite flight level air temperature in altitude correction window. Opposite flight level pressure altitude on inner scale, find true altitude above sea level on outer. Subtract ground elevation from true altitude to find absolute altitude.

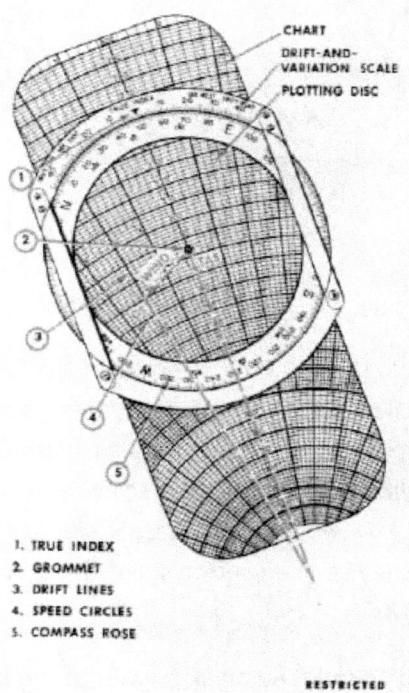

Figure 4. Plotting surface on front side of the E–6B computer

Vector solutions. The front side of the E-6B computer (Figure 4) consists of the transparent plotting surface, the sliding chart, and the drift-and-variation scale. This side is used to solve wind vectors and similar problems without plotting the complete triangles. Otherwise you solve vector problems exactly as you would if plotting them on graph paper. Each of the three sides of the triangle is called a vector. The vectors are:

Direction	Wind direction	True heading	True course
Speed	Wind speed	True airspeed	Groundspeed

If you know any 2 of the 3 directions and any 2 of the 3 speeds you may solve for the other direction and speed on the computer. Plot the wind arrow on plotting disk by setting wind direction at true index and tracing measurement of wind speed from grommet down centerline of chart. Then, mark end with short crosswise line and/or point. Plot drift on plotting disc tracing along appropriate radiating drift line. Plot groundspeed on plotting disc by tracing along arc of appropriate groundspeed circle. Note. When 3 drift lines intersect, or form a small triangle, you can be reasonably sure you have an accurate wind solution. When you solve for true course properly, your true course is at your drift on drift-and-variation scale when point of arrow is that of your drift on drift lines."

MIT Professor Vannevar Bush

Vannevar Bush (1890-1974) was an American engineer, inventor and science administrator, who during World War 2 headed the U.S. Office of Scientific Research and Development (OSRD), through which almost all wartime military R&D was carried out. His first name appears to be unique. He was named after a friend of the family "John Van Nevar." In 1919 Bush joined the MIT Department of Electrical Engineering. Bush and others of MIT's electrical engineering staff, discouraged by the time-consuming mathematical computations required to solve certain engineering problems, began work on a computing machine that would solve differential equations. In 1927, they started construction of a differential analyzer, which was an analog computer that could solve differential equations. As time went on, they built improved versions. The differential analyzer had a table-like array of shafts and pens that mechanically simulated and plotted the fina results. Unlike earlier designs that were purely mechanical, the differential analyzer was an electromechanical device with both electrical and mechanical components. In 1930 Karl T. Compton was appointed president of MIT. Compton appointed Vannevar Bush to the newly created post of MIT vice president in 1932. Also Bush became the dean of the MIT School of Engineering.

In 1935, Bush unveiled the second version of the differential analyzer. It filled a room with a complicated array of gears and shafts driven by electric motors. It weighed 100 tons. At the time, it was marvelous. By present standards the machine was slow, only about 100 times faster than a human operator using a desk calculator. It was cumbersome, clumsy, and massive. In addition to all of the mechanical elements (the integrators, torque amplifiers, drive belts, shafts, and gears), the differential analyzer contained 2000 vacuum tubes, thousands of relays, 150 motors, and approximately 200 miles of wires connecting relays and vacuum tubes.

The differential analyzer was based upon the measurement of mechanical movements and distances. It used shaft movement to represent variables, gears to multiply and divide, and differential gears to add and subtract. Integration was done by a sharply-edged wheel spinning at variable radius on a round rotating table. To provide amplification, the differential analyzer employed torque amplifiers which were based on the same principle as a ship's capstan.

The machine was designed to solve differential equations by integration. In simple terms, the differential analyzer found the area under a given curve. Six wheel-and-disc "integrators" at its heart could be connected to 18 long, rotating shafts. These integrators had to be arranged in a specific way for each particular type of differential equation. The wheel-and-disc mechanism performs integration in essentially the same way as the milometer on an automobile gives the cumulative mileage. In a differential analyzer, the output of one integrator drove the input of the next integrator, or a graphing output. The machine could solve, approximately, an arbitrary sixth-order differential equation. However it had to be laboriously set up for each new problem. The differential analyzer was the evolutionary jump from mechanical to electrical and electronic components in computers. The similarity between linear mechanical components, such as springs and dashpots (viscous-fluid dampers), and electrical components, such as capacitors, inductors, and resistors is striking in terms of mathematics. They can be modeled using equations of the same form.

In World War 1, the Germans developed the huge railway gun. On testing, the shell went twice as far as expected. Their ballistic tables had not taken into account that drag on the shell was significantly reduced at the high altitudes reached by the trajectory. From that time on, computation of exacting ballistic tables became essential. In Maryland, at Aberdeen Proving Ground, the U.S. Army Ordnance Department computed numerical tables of ballistic trajectories. These calculations were an attempt to mathematically model every possible field condition while taking into account the weight and shape of the shell and its propellant charge. The data were compiled into the firing tables that field gunners consulted in order to aim their weapons.

In 1914, the first commercial desk calculators began entering businesses and the use of calculators began to be popular. At first the laborious calculations required for firing tables were done on desk calculators. In 1934 the Moore School of Electrical Engineering at the University of Pennsylvania constructed two differential analyzers, one on campus and one at Aberdeen. All was done with the full cooperation of MIT and Vannevar Bush. During World War 2, these two differential analyzers, as well as the ones at MIT, were used to compute ballistic tables. MIT had the domain of computing securely within its grasp. In 1945 it appeared that MIT's position of preeminence in computing could not be challenged.

Chapter 6. Digital computers

Alan Turing: A computer would deserve to be called intelligent if it could deceive a human into believing that it was human.

Stonehenge as a digital calculator

Exploration is the act of searching for the purpose of discovery. Early peoples used their eyes to explore the heavens. They surveyed the stars and observed the movements of the sun and moon. In the deserts of the Middle East are huge ancient stone circles with diameters of up to 400 meters. Prehistoric stone circles are also found throughout Europe. In Britain and Ireland Stone there are hundreds of stone circles. Stonehenge is the most impressive structure surviving from Neolithic Europe. Stonehenge was designed as a huge display that exhibited wonders of the heavens; in particular wonders that were uniquely associated with the latitude of Stonehenge. A stone circle north of Stonehenge or one south of Stonehenge would not possess these wonders. In this respect Stonehenge is unique.

Stonehenge underwent three major phases. We are concerned with the first monument, known as Stonehenge 1. It consisted of a circular bank and ditch, with a large entrance to the northeast. Stonehenge 1 is dated to around 3100 BC. Within the outer edge of the enclosed area is a circle of 56 holes dug into the surface chalk. The diameter of the circle is 288 feet. Each hole is about 3 feet in diameter. They are known as the Aubrey holes, named in the seventeenth century after John Aubrey who identified them. Most likely the holes contained standing timbers. It is also possible that the holes held bluestones that were later removed for other uses.

The *summer solstice* (otherwise known as midsummer or the first day of summer) occurs when the sun reaches its highest point in the sky at noon. It is the longest day of the year and takes place about June 20. (called midsummer or the first day of summer). The *winter solstice* (otherwise known as midwinter or the first day of winter) occurs when the sun reaches its lowest point in the sky at noon. It is the shortest day of the year and takes place about December 20. The equinox occurs twice each year. It is when daylight and night darkness are of equal length. The *autumn equinox* (otherwise known as the first day of fall) takes place about September 22. The *spring equinox* (otherwise known as the first day of spring) takes place about March 20.

On what knowledge did the Britons build Stonehenge 1? The Britons in studying the movements of the sun and moon discovered some amazing facts (or wonders). The reason for these wonders is that Stonehenge is located at latitude of 51°. Stonehenge was built to demonstrate these facts. The altar is at the center of the circular bank and ditch. The Britons observed the rising sun at Stonehenge over many years. The sun rises almost due east at spring equinox. The sun also rises almost due east at fall equinox. The sun rises farthest to the northeast on summer solstice. It rises farthest to the southeast on winter solstice.

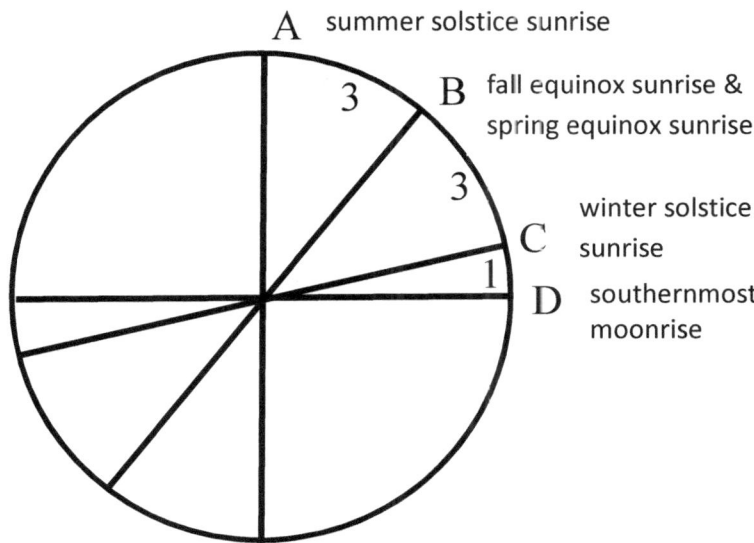

Figure 5. Stonehenge 1 is dated to around 3100 BC. It exhibits the numbers 3, 3, 1, which add up to the magic number 7 for a right angle.

The ancient Britons would sight the rising sun from the center of the circle. On the following occasions they dug a hole at the point where the line of sight crossed the circle (Figure 5).

> At the summer solstice sunrise, they dug hole A.
> At the fall equinox sunrise, they dug hole B.
> At the winter solstice sunrise, they dug hole C.
> At the spring equinox sunrise, it was hole B again.
> At the southernmost moonrise they dig hole D.

Like a pendulum, the sunrise direction changes from northeast at summer solstice to southeast at winter solstice, and back to northeast, completing one full swing back and forth, or one full cycle, each year. There is a stopping or standstill at the extremes, or solstices. The word solstice means "standstill of Sun." The term "lunar standstill" describes the similar extremes in the moon's directions. In other words, lunar standstill (for the moon) is the counterpart of solstice (for the sun.)

The directions of moonrise and moonset change like a pendulum, swinging back and forth along the horizon and completing one full swing each month. At MAJOR STANDSTILL, the Moon exhibits the maximum monthly range of rising and setting directions. At the latitude of Stonehenge, the angle between the winter solstice sunrise directions and southernmost moonrise direction is exactly 90°. The reason is that Stonehenge lies on the exact latitude at which the midsummer sunrise and sunsets are at 90° of the moon's northernmost setting and southernmost rising. This particular phenomenon is only possible within a band of less than one degree, of which Stonehenge lies in the middle-third. In summary, the hole locations and sightlines indicate summer solstice sunrise A, equinox sunrise B, winter solstice sunrise C, and southernmost moonrise D.

The Britons took the summer solstice sunrise direction as the reference line for Stonehenge. They walked along the circle and measured the distance between the four stones. They found that the distance between A and B was the same as the distance between B and C. They also found that this common distance was three times the distance between C and D. In other word they found that

   a. The distance between A and B was 3
   b. The distance between B and C was 3
   c. The distance between C and D was 1

These three distances add up to 7, and 7 is a magic number. This 3, 3, 1 relationship is only true at the latitude of Stonehenge. This numeric magic is the wonder of Stonehenge. They built Stonehenge as an analog device that would demonstrate this wonder. They filled in four more stones so that the distance between any two adjacent stones would be same as the distance between C and D. Also they realized that the distance between A and D was one quarter of the circle. They filled in the rest of the circle. The total number of stones for quarter circle (quadrant) was 4 X 7=28. (Note that 7 is the number of days in a week and that 28 is approximately the number of days in a month.) For good measure, they placed other stones halfway between existing stones. Thus they obtained 2 X 28 = 56 stones in total. The entire circle is made up four quadrants; that is, 4 times 14, which is the desired number 56.

   a. Aubrey hole 0 is on the summer solstice A.
   b. Aubrey hole 6 is on the equinox B.
   c. Aubrey hole 12 is on the winter solstice C.
   d. Aubrey hole 14 is on the right angles line D.
   e. Aubrey hole 56 is the same as hole 0.

Gerald Hawkins was a British-born American astronomer and author noted for his work in the field of archaeo-astronomy. He proposed that the Aubrey holes were used for calculating the phases of the moon and also for predicting the month of the year in which eclipses would take place. In other words, Prof. Hawkins made the point that the Aubrey circle could be used as a numerical counting device (a digital calculator).

Blaise Pascal and Gottfried Wilhelm Leibniz

Blaise Pascal (1623-1662) invented the mechanical digital calculator. Pascal started on the project in 1642 when he was 19, and three years later he had a working machine. His calculating machine, named the *Pascaline*, could add and subtract two numbers directly and multiply and divide by repetition. The device would carry digits from one column to the next.

Different monetary systems, based on the Latin words *librae*, *solidi*, and *denarii*, were widely used in Europe in medieval times. The French system used livres, sols and deniers. There were 20 sols to a livre and 12 deniers to a sol. The English system used pounds, shillings, and pence. There were 20 shillings to a pound, and 12 pence to a shilling. In France, length was measured in toises, pieds, pouces and lignes. There were 6 pieds to a toise, 12 pouces to a pied, and 12 lignes to a pouce. In England, length was measured in fathoms, feet, inches, and lines. There were 6 feet to a fathom, 12 inches to a foot, and 12 lines to an inch. In addition to wheels in base 10, the Pascaline needed wheels in base 6 (for pieds to toises), in base 12 (for deniers to sols, for pouces to pieds, and for lignes to pouces), and in base 20 (for sols to livres).

With the help of local craftsmen, Pascal worked on the idea. Many experimental models were constructed and by 1645 he had a working machine. It was the size of a glove box. By 1652 he had a standard model in production. It was priced 100 livres. Unfortunately his hopes to get rich were not realized. Previous machines were analog. Blaise Pascal had invented the first working mechanical digital calculator, a major innovation whose descendants were to change the world. Pascal wrote:

> We owe a great debt to those who point out faults. For they mortify us. They teach us that we have been despised. They do not prevent our being so in the future; for we have many other faults for which we may be despised. They prepare for us the exercise of correction and freedom from fault.

Gottfried Wilhelm Leibniz (1646-1716) is best known for calculus, but he also constructed the *Step Reckoner*, a device which, as well as performing additions and subtractions, could multiply, divide, and evaluate square roots by series of stepped additions. Charles Xavier

Thomas de Colmar (1785-1870) patented the first practically successful calculator in 1820. His machine worked on the stepped gear principle of Leibniz. Derivative devices were still being manufactured after 1900 and some were still in use in the 1940s.

Charles Babbage and Ada Byron Lovelace

Calculators are designed with specific functions such as addition, multiplication, and logarithms. In 1822 Charles Babbage (1791-1871) conceived the *Difference Engine*. Its purpose was to calculate mathematical tables automatically. The *Difference Engine* was only partially completed when Babbage conceived the idea of a more sophisticated machine called the *Analytical Engine*. It was a major innovation. This machine was a universal calculating engine, a machine that would perform any series of calculations that could be formulated for it. The *Analytical Engine* was not a logical extension of the *Difference Engine* concept, but radically different. In it Babbage foresaw all the major components and functions of a modern digital computer.

The French inventor Jacquard had worked out a method for weaving patterns in rugs as determined by the use of punched cards. The *Analytical Engine* was intended to use loops of Jacquard's punched cards to control the operations. The cards would not always be fed into the machine in constant succession one after the other. In order to perform a program loop, sequences of cards could be reused. Depending on the result of a calculation, a conditional jump could be made.

Unfortunately the *Analytical Engine* was never completed. It was not a single physical machine. Instead it was a succession of designs that Babbage made until his death in 1871. It may be said that the *Analytical Engine* represented the first mechanical digital computer. The fundamental difference between a digital calculator and digital computer is that a computer can be programmed in a way that allows the program to take different branches according to intermediate results.

Ada Byron Lovelace was the daughter of the poet Lord Byron. She is distinguished by the fact that she was the first person to foresee the universality of the digital computer. Using the *Analytical Engine* as a model, Ada Lovelace devised the concept of a programmable digital computer. Babbage focused only on the capabilities of a machine to do routine arithmetical computations. Ada Lovelace had the genius to foresee the capability of a digital computer to extend beyond numerical calculations. She showed that the potential of a digital computer went into regions far removed from what anyone else had ever realized. Ada Lovelace wrote:

> The *Analytical Engine* might act upon other things besides number, if objects were found whose mutual fundamental relations could be expressed by those of the abstract science of operations. Supposing, for instance, that the fundamental relations of pitched sounds in the science of harmony and of musical composition were susceptible of such expression and adaptations, the engine might compose elaborate and scientific pieces of music of any degree of complexity or extent. We may say most aptly that the *Analytical Engine* weaves algebraic patterns just as the Jacquard loom weaves flowers and leaves.

The ENIAC

Electronics made the modern computer possible. The Pascaline was the first working mechanical digital calculator. The ENIAC was the first working electronic digital computer. Work was begun on the Pascaline in 1642; on the ENIAC in 1943. It was a time span of 301 years. The Pascaline and the ENIAC are the two breakthroughs that changed the world from analog to digital. During World War 2, engineers at the Moore School of the University of Pennsylvania decided to design a computer that could do numerical integration (NI) more rapidly than the analog integration done on the differential analyzer. The proposed machine would be electronic; that is, it would be an Electronic Numerical Integrator (ENI). And then someone added that it would be even more general still, and so the words "And Computer" were added.

The machine was named the *Electronic Numerical Integrator And Computer*. Its acronym was ENIAC.

When completed in 1946, the ENIAC was moved from the Moore School to Aberdeen Proving Ground. **Dr. Joseph H. Levin** was the Chief of the Machines Branch at Aberdeen. In this capacity he was in charge of its differential analyzer. Would the ENIAC displace the differential analyzer? Levin fought the good fight. He wrote the classic paper (Levin, 1948) which carried the differential analyzer to its greatest attainment.

The ENIAC was digital, but it was not a stored-program computer. The ENIAC was essentially a huge pile of parts. Signals were sent among the components by cables (wires), which were plugged in by hand. To solve a mathematical problem, components of the ENIAC would have to be assembled, mainly through wiring, into a special purpose calculator. After serving in the Army during World War 2, **Dr. Bernard Dimsdale** took a position at Aberdeen early in 1947. On his first day he was handed a nonlinear partial differential equation and a stack of wiring diagrams about a foot thick. Dimsdale was told to compute the solution of the equation. Dimsdale fought the good fight. He spent weeks on the ENIAC making wiring connections, and he set hundreds of switches. He connected input and output terminals so as to form digital trunks for the communication of numerical data. The units had to be set up to recognize when they were to operate and which particular operations were to be performed. There were no guidelines available, so everything was a major undertaking. When he threw the power switch on, the vacuum tubes blew out in large batches. The ENIAC had no power control device. Tubes were replaced time and again. He spent some more weeks in rewiring. No luck. The ENIAC was temperamental; it was a nervous machine. Dimsdale told everyone to be careful around it, and never to lean on anything. In time he found whole rows of unsoldered connections within the components. The output was the card punch, and from its chatter he could tell how things were going. He found that the machine was effective only late at night when most of the electric lights in the neighborhood were turned off. Only a perfectionist like Dimsdale had any chance at all. After some more attempts, he finally got the ENIAC to work. It seemed like a miracle. In a few minutes the machine ground out the required answers for this particular mathematical equation. The next mathematical problem would require a completely new assembly from scratch. It was clear that the ENIAC as designed was a white elephant.

Dr. Levin used the analog differential analyzer. Dr. Dimsdale used the digital ENIAC. They were the heavyweights in computing. Would it be analog or digital? **Dr. Richard F. Clippinger**, the head of the Mathematics Division at Aberdeen, stepped in. In 1947 he set out to modify the ENIAC into a stored program computer. He had electronic engineers from Aberdeen permanently assemble the machine into a fixed configuration, a process that took some months. In essence, Clippinger took the assortment of parts that made up the original machine and from these parts built a stored-program computer. For four days a year, John von Neumann was a consultant to Aberdeen Proving Ground. Clippinger and Dimsdale would meet with him on these occasions and they would fill him in on the modifications they were making on the ENIAC. By the summer of 1948 Clippinger had succeeded in converting the ENIAC into the world's first stored-program digital computer with a programming language (Clippinger, 1948). The programs were fed into the machine on the function tables (the banks of switch-controlled resistor matrices originally designed to hold input data).

The new method worked beautifully, but others wanted to go back to the original way. They said the ENIAC was at least six times faster with the old way. But Dimsdale came to the rescue. He said although that is true, it is not important. The old way requires some months to configure the ENIAC by rewiring the machine, and then a few minutes to do the calculations. The ENIAC could not be used by others during the periods of rewiring. The new way requires a few days to write the code, and a few hours to do the calculations. People from all over could write codes, and all of the codes could be run on the computer, thereby significantly multiplying the output of the computer. The old way of rewiring the machine for each problem was never used again. Instead a code would be written in the programming language for each problem. Over its lifetime the ENIAC did a considerable amount of computing on all sorts of projects.

The IEEE Computer Society presents its *Computer Pioneer Award* to a person who made significant contributions to concepts and developments in the electronic computer field which have clearly advanced the state of the art in computing. The contributions must have taken place fifteen or more years earlier. In 1996 Richard Clippinger received the *Computer Pioneer Award* for his conversion of the ENIAC to a stored program computer.

Modern computer history begins with the ENIAC. Until about 1952 the ENIAC was the main computer in the world for the solution of the scientific problems. The ENIAC logged in a total of 80,223 hours of operation in its lifetime from 1946 to 1955. In addition to ballistics, the ENIAC was used for other scientific endeavors including atomic energy calculations, weather prediction, wind tunnel design, cosmic ray studies, thermal ignition, and pseudo-random number generation. The ENIAC is reputed to have done more arithmetic than the entire human race had done prior to its construction. Included in this previous arithmetic were the additions and multiplications performed by every school child in history.

In the summer of 1949, I attended the ROTC summer camp at Aberdeen Proving Ground in Maryland and saw the ENIAC. The ENIAC was made up of thirty units, each one eight feet tall, three feet deep, and from two to six feet wide. Lined up side by side, they stretched for about ninety feet in a U along three sides of the room. The machine had 18,000 vacuum tubes, 70,000 resistors, and 6,000 switches, and consumed 140 kilowatts of power. The ENIAC included a cycling unit, twenty accumulators, an initiating unit, a high-speed multiplier, a combined division and square root unit, function tables, and input and output units. The function tables were made up of three panels. Data as well as instructions were entered on the function tables by dial switches which were set by hand. These switches selected the digits and signs for each of the 104 values of an independent variable that were stored in each table. In addition to the function tables, there were other ways of supplying the machine with information (data or instructions). The requisite numbers could be put into the machine by means of punch cards fed into the card reader, or by means of switches on the constant transmitter.

Whirlwind

At the end of World War 2, MIT made plans to retain its preeminent position as the world's center of computing. A whole new stable of analog machines was planned around the differential analyzer. But surprisingly quickly the effort collapsed. In 1950 the MIT provost confided to the MIT president that MIT has missed its chance to retain leadership in computing. And what was the reason? It was the ENIAC, and the host of digital computers that it spawned: the EDVAC, the ORDVAC, the UNIVAC, the Institute of Advanced Study computer, the JOHNIAC, the MANIAC, the ILLIAC, the SILLIAC, the EDSAC, the Manchester Mark 1, and looming up, right at MIT, the Whirlwind.

What was the Whirlwind computer? With the differential analyzer in mind, the government in 1943 gave MIT a contract to build a flight training simulator. It was called Whirlwind. Jay W. Forrester was the director of the Project, and the assistant director, Robert Everett, later became a founding father of the MITRE Corporation. For a real-time simulator, speed was essential. The proposed analog simulator was appropriately called Whirlwind. However an analog device turned out to be too slow to calculate the responses to a pilot's actions. In desperation Forrester went to see the ENIAC in November 1945. The result was that Forrester changed Whirlwind into a digital computer.

At the MIT Radiation Laboratory in Building 20, Prof. Valley had developed the directed radar that played a strategic role in World War 2. After the war he worked on the air defense of the United States using the radar technology. On a visit to Whirlwind in 1950 Prof. Valley realized that the successful air defense would require the linking of computers and communication. This idea was the germ that grew into the Internet. The U.S. Air Force took over the funding of Whirlwind, which became fully operational in 1951. Whirlwind was the prototype for a new system called SAGE that would provide computerized electronic defense against an air attack.

The Barta Building behind the MIT campus was the home for Project Whirlwind. The building was linked by telephone lines to radar sites. The computer occupied 2500 square feet on the second floor. The MIT Digital Computer Laboratory with its programming and maintenance personnel had offices on the first floor. Whirlwind was a single-address binary computer. It had 4,500 vacuum tubes and 14,800 diodes. Because short words helped real-time operations, Whirlwind had a 16-bit word structure instead of the longer formats geared for scientific computations. Initially the high-speed memory (RAM) was electrostatic storage with a size of 1,024 words. Within a few years the electrostatic storage was replaced by magnetic core memory, and the size was increased. Initially the add time was 20,000 additions per second (0.05 milliseconds per addition), but it was increased to 50,000 additions per second. Power supplies were located in the building's basement and the roof was covered with air conditioning equipment to cool down the system.

Whirlwind in its greatly improved version became the real-time computer that was at the heart of the SAGE communications and control network. It had leapfrogged all of the other computers that had been spawned by the ENIAC. The stored-program ENIAC was the invention that brought forth digital computing as the confluence of the two streams: calculation and programming. Whirlwind was the invention that brought forth the confluence of two streams: communication and computing. MIT had regained leadership, but now it was in communication and computing.

The Semi-Automatic Ground Environment (SAGE) was a continental air defense system that exceeded the Manhattan project in cost and scale. IBM built the computers, Burroughs developed communications, Western Electric constructed the concrete "Direction Center" buildings, and MIT (and, after 1958, the MIT-formed not-for-profit corporation called MITRE) provided system integration. By the time the SAGE system was fully deployed in 1963, the 23 Direction Centers and three Combat Centers were linked by long-distance telephone lines and radio contact to more than 100 interoperating air defense elements. At the heart of SAGE was the giant Whirlwind II (AN/FSQ-7) computer. Each of the Direction Centers was equipped with two Whirlwind II computers: one operating live and one operating in standby mode for additional reliability. SAGE required system integration on a scale previously unimagined. Whirlwind II ran the largest computer program written up to that time, with 500,000 lines of code. SAGE employed digitized radar data, long distance data communications via landlines and ground-to-air radio links, and featured a large collection of interactive display terminals. The program automated information flow, processed and presented data, and provided control information to the weapons systems.

The communications devices from Burroughs allowed each center to communicate with other centers, creating the first large-scale computer network. SAGE was responsible for training more than 10,000 computer programmers in the 1950s, and many later worked for the Advanced Research Projects Agency (ARPA). SAGE pioneered the important technology which is used to facilitate Internet processing today; for example, the modem, the mouse in the form of a light gun, multi-tasking, array processing, computer learning, fault detection, and interactive computer graphics. When this technology was transported to ARPA, the result was ARPANET. The Internet sprang from ARPANET. The last of the Whirlwind II computers shut down in 1983, giving Whirlwind the record for practical operational longevity among all digital computers. But in a larger sense, Whirlwind never did shut down; its technology drives the Internet.

Chapter 7. Cybernetics

Norbert Wiener: The nervous system and the automatic machine are fundamentally alike in that they are devices, which make decisions on the basis of decisions they made in the past.

Thermometer and thermostat

In winter a house needs to be continually heated. Loss of heat occurs through doors, windows, and even through walls. A thermometer is a device that measures temperature. When the room is warm, the thermometer on the wall has a high reading. When the room is cold, the thermometer on the wall has a low reading. A thermometer has no link to the heater. The thermometer cannot make the room warmer or colder. The heater affects the thermometer. However the thermometer does not affect the heater. A thermometer and heater separately do not represent a feedback loop.

Now let the instrument on the wall not be a thermometer but a thermostat. A thermostat links thermometer and heater. On a thermostat you set the desired temperature. It is the setpoint. Thermostats use a thermometer to measure the room temperature. The thermostat turns the heater on when the room temperature falls to a certain point below the setpoint. The heater runs at full capacity until the setpoint is reached, and then the heater shuts off. A thermostat is a bang-bang feedback system in that the heater (when on) always runs at full capacity. In an ordinary feedback system, the heater would run in an amount proportional to the difference between room temperature and the setpoint.

A thermostat is a feedback control system that maintains the room temperature close to a desired level. Not only does the room temperature depend upon the action of the heater, but also the action of the heater depends on the room temperature. One causes the other and vice versa. In other words, there is a feedback loop.

Not all automatic control systems use feedback. An example is a purely feedforward system. Let us give an example. The thermometer is placed in the open air outside the house. The outside thermometer is connected so as to operate the fuel valve. In other words, the outside temperature, and not the inside temperature, regulates the fuel flow. In this feedforward system, the room temperature has no effect on the fuel flow; there is no feedback loop.

Such a feedforward system is not necessarily a useless system; it does find some application. A woodsman is a person accustomed to life in the woods and skilled in the arts of the forest, as hunting or trapping. In the evening, the woodsman looks at his thermometer on a tree to see what the outside temperature is. In his mind, the woodsman has calibrated the proper supply of fuel for any given outside temperature. He accordingly brings into the log cabin a certain amount of wood to burn that night. The colder the outside temperature is, the greater the amount of wood. He had no thermometer in the cabin, so he does not know the room temperature. However, this method can only deal with standard conditions. On a windy day there will be more heat loss, so the inside temperature will be less than what the woodman would like. A feedback system circumvents such limitations. Feedback directly addresses the quantity to be controlled, and it comprehensively corrects for all kinds of disturbances. It does not require calibration for each special condition as does a feedforward system.

Governor for the steam engine

In ancient times, the float valve was used to regulate the flow of water in Greek and Roman water clocks; similar float valves are used to regulate tank water level in the flush toilet. In the 17th century, centrifugal governors were developed to regulate the distance and pressure between millstones in windmills. Early steam engines with purely reciprocating motion were used for pumping water. They could tolerate variations in the working speed so governors were not needed. The Scottish engineer James Watt developed the rotative steam engine to operate factory machinery. Constant operating speed was obtained by Watt's fly-ball governor. It was a centrifugal feedback valve for controlling the speed of the engine. It used weights mounted on spring-loaded arms to determine how fast a shaft is spinning, and then uses proportional control to regulate the shaft speed. The theoretical basis for the operation of governors was described by James Clerk Maxwell in 1868 in his seminal paper "On Governors." The term governor was used for any device that automatically regulates the supply of fuel, steam, or water to a machine, ensuring uniform motion or limiting speed. In the twentieth century governors found all kinds of uses in communication and control. In time, the word governor was replaced by the word "servo-mechanism" or "servo" for short.

Steam engine governors fall into various classes. Here we will describe the throttling governor, which throttles the steam in the supply pipe. The throttling governor consists of a throttle valve placed on the steam pipe. This valve is attached to a spindle. At the upper end of the spindle are the two fly-balls. The spindle and fly-balls form what is known as a revolving pendulum. The spindle and balls are driven from the main shaft by a belt and bevel wheels.

Suppose the engine moves faster than the desired speed. In such a case, the fly-balls are forced to revolve at a higher speed, and will, consequently, move outwards and upwards through the action of centrifugal force. This action forces the spindle downwards, and partly closes the throttle valve. The engine thus takes less steam, and the speed falls to the desired point, the governor balls in the meantime returning to their original position. On the other hand, suppose the engine moves slower than the desired speed. The fly-balls drop and open the valve wider. More steam is admitted, and the engine regains its original speed.

Let us summarize. It is desired that the engine run at a target speed. The target speed is the signal that is communicated to the engine by the engineer. The error signal is the departure of the actual speed from the target speed. This error is interpreted by the governor to adjust the throttle. The resulting change reduces the error in speed. The terms positive and negative feedback may be defined as the altering of the error between target value and actual value of a parameter. The parameter in this case is speed. If the error is widening, it is called positive feedback. If the error is narrowing it is called negative feedback. The steam engine governor makes the error narrow, so it is an example of negative feedback.

Governor for the heart

Physiological parameters must remain within a narrow range for a person to survive. Negative feedback loops within the body keep physiological parameters such as heart rate within the target range. For example, the average resting heart rate should remain in the target range from 60 to 100 beats per minute. A negative feedback loop works by adjusting an output, such as heart rate, in response to a change in input, such as blood pressure. A basic loop consists of a receptor, a control center and an effector. In this case, the receptor measures blood pressure, the control center is the brain, effector is the heart muscle.

If a person is at rest and blood pressure increases, pressure receptors in the carotid arteries detect this change in input and send nerve impulses to the medulla of the brain. This action tells the brain to reduce nerve impulses that stimulate the heart muscle to contract. The heart contracts more slowly and the heart rate decreases, causing blood pressure to decrease to within target levels.

When exercising there is an increase in heart rate and blood pressure. The body increases blood flow to muscle tissue in response to the increased demand for oxygen. The homeostatic set points of heart rate and blood pressure are therefore "reset" higher. Vigorous exercise can make heart rate increase to as much as 180 beats per minute or more. Negative feedback loops act to maintain heart rate and blood pressure within these new higher target ranges. After exercise the muscle tissues no longer demand as much oxygen so homeostatic set points are reset back to the original target.

Servomechanisms

A modern servomechanism is an electronic system in which a hydraulic, pneumatic, or other type of controlling mechanism is actuated and controlled by a low-energy signal. A servomechanism acts continually on the basis of information to attain a specified goal in the face of changes.

In the early days the rudder of a ship was directly connected to the tiller, as it is today on small sailboats. However, appreciable power amplification is needed when we consider the massive size of the rudders of large ships and the forces which must be exerted to move them. The system that automatically controls the position of the rudder of a large ship is a servomechanism. In controlling the heading of the ship the helmsman exerts a comparatively small amount of torque about the axis of the helmsman's wheel. The servomechanism positions the rudder by use of a source of power. The position of the helm is the input to the servo and the position of the rudder is the output from the servo.

Power steering systems for cars augment the steering effort of the driver by means of an actuator. An actuator adds controlled energy to the steering mechanism. As a result, it takes less physical effort for the driver to turn the wheels when driving at speed. Also it takes less physical effort for the driver to turn the wheels when a vehicle is stopped or moving slowly. The power system has a direct mechanical connection between the steering wheel and the linkage that steers the wheels. In case the power-steering system fails to augment the physical effort, the driver can still steer the vehicle by using manual effort alone.

In brief, power steering is achieved by a servomechanism. It is an automatic feedback control system in which the motion of the output member (e.g. the wheels of the vehicle) is constrained to follow closely the motion of the input member (the steering wheel), and in which power amplification is incorporated. The motion of the steering wheel is the cause and the motion of the wheels is the effect. More generally, in any servo system there is a cause-effect relationship, namely, the input is the cause and the output is the effect. The term "servo" is derived from a Greek word which means slave. The servomechanism behaves in a slave-like manner. The output member is constrained to repeat the motion of an input member at an increased power level.

MIT Professor Norbert Wiener

Professor Norbert Wiener (1894–1964) was a most esteemed member of the MIT Mathematics Department. At 11 years of age, Norbert Wiener entered Tufts College. He was awarded a BA in mathematics in 1909 at the age of 14. Harvard awarded Wiener a PhD in 1912, when he was 17 years old. His dissertation was on mathematical logic. He showed that ordered pairs can be defined in terms of elementary set theory. It follows that relations can be defined by set theory, so the theory of relations does not require any axioms or primitive notions distinct from those of set theory. Wiener served in the Army in World War 1. He was stationed at Aberdeen Proving Ground in Maryland and he worked on ballistics. He was honorably discharged from the Army in February 1919. Wiener became a member of the MIT Mathematics Department, where he spent the remainder of his career.

Wiener is famed as pure mathematician, but it was his uncanny ability to apply mathematics to an extensive range of applications that made him unique. He was friendly and open. He worked well with people and was never afraid to tackle new things. Wiener had an interest in medicine and physiology and sought out people in these areas. He was an early researcher in stochastic and noise processes, contributing work relevant to electronic engineering, electronic communication, and control systems. In the 1930s, he consulted with Vannevar Bush on the development of the differential analyzer. In 1940 Vannevar Bush assisted in the formation of the National Defense Research Committee (NDRC). It brought together university people and industrial people to work on military research. Section D-2 worked on control systems.

The expression "fire control" has two different meanings. One meaning is "the prevention and monitoring of forest fires and grass fires." The other meaning is the one we use here; namely, "the process of targeting and firing heavy weapons." The Army and Navy were especially interested in the best way to aim their long range guns. The NDRC asked

Wiener to bring his knowledge of networks to antiaircraft fire control. Almost all of the antiaircraft guns used by the Army in World War 2 were manufactured at Watertown Arsenal, which was located a few miles from MIT. Fast-moving airplanes were difficult to hit. The key element in fire control was prediction. The shell's time of flight could be as much as one minute for high flying aircraft. Ways were needed to predict a moving target's position with high precision despite corrupted data and the failures of human operation. On December 1, 1940, NDRC Section D-2 gave a contract to MIT for "General Mathematical Theory of Prediction and Applications." This contract funded Wiener's work in the MIT Mathematics Department. Wiener and his assistant, engineer Julian Bigelow, would devise a way to follow the path of an airplane, and then estimate the future course of that path.

During early 1941 Wiener and Bigeow designed and built a machine to simulate their idea on the automatic aiming and firing of anti-aircraft guns. Wiener and Bigelow quickly ran into a stability problem. Their machine was highly sensitive, even unstable, in the presence of high frequency noise. The noise caused oscillations that prevented accurate aiming. Wiener quickly realized the problem was fundamental so that he would need a new approach. Wiener now turned to statistics, designing a new predictor based on statistical analysis. More specifically, the predictor was based upon autocorrelation, which connects past and future.

Through the remainder of 1941, Wiener worked out the mathematical details. Building on his own work in harmonic analysis and operational calculus, Wiener constructed a general theory of smoothing and predicting time series. Wiener prediction theory is linear prediction. Often people use the term "linear prediction" instead of "Wiener prediction." Wiener worked mainly with the case of continuous time series (i.e., analog signals) but did include brief summaries for the case of discrete time series (i.e., digital signals). Wiener gave his theoretical results in "Progress Report to the Services," No. 19, *Extrapolation, Interpolation, and Smoothing of Stationary Time Series with Engineering*

Applications. It was published by the NDRC for restricted circulation in February 1942.

Wiener found himself embedded in an analog environment. Antiaircraft fire control has to be done in a split second on the spot in real time. Wiener's prediction methods made use of data on the aircraft's flight path. The equations would have to be solved almost instantaneously. The 100-ton MIT electromechanical differential analyzer was unable to do the task. The sad fact is that Wiener's methods were too advanced to be implemented on existing analog equipment. As a consequence, Wiener's work on prediction was largely unappreciated.

The war effort needed people that worked on the solution of immediate problems with the resources at hand. A dedicated engineer appeared to be more qualified for such work than a theoretical mathematician. Unfortunately, in January 1943 Section D-2 did not renew Norbert Wiener's contract. No consideration was given to the untold advantages of having the genius of Wiener working on these problems. No consideration was given to the great psychological incentive that Norbert Wiener imparted to coworkers. The development of Wiener's prediction theory in engineering was discontinued for the rest of World War 2. However, in 1942 Professor George Wadsworth in the Mathematics Department took up Wiener's work and used it for numerical weather forecasting. This story will be told in Chapter 11.

As an analogy, the ancient Greeks through mechanical ingenuity and skill developed the feedback controlled water clock. Centuries later, Christiaan Huygens through mathematics developed the feed-back controlled pendulum clock which kept much more accurate time. During World War 2, engineers through electromechanical ingenuity and skill developed intricate servomechanisms. Norbert Wiener through mathematics developed intricate scientific systems which could not be implemented at the time. Today with digital technology Wiener's work is coming into fulfillment.

The cancellation of his NDRC contract in 1943 put Wiener outside the massive wartime research effort. Wiener was not one to rest upon his laurels. No longer would Wiener have access to classified engineering developments. Wiener would only have access to civilian resources. Wiener plunged into furthering his work in a completely different and separate direction. Free of the NDRC's engineering demands, Wiener was able to choose his own course of action. It would along more esoteric lines. Wiener broke the ground for an entirely new area of research: the use of engineering methods in biology and medicine. It has had a far reaching effect. Today MIT is doing extensive research and development along these lines with great success.

In 1943 Wiener started important research in medicine and physiology. He looked into physiological and neurological feedback systems that governed life's functions. In so doing he developed the idea of cybernetics which explores the biological feedback mechanisms. He was already well qualified in the area as he was regularly interacting with researchers in neuropsychology and biophysics. To Wiener the core of cybernetics is the servo-mechanical nature of the mechanisms of control and communication in both humans and machines. His program sought to extend that understanding to biological, physiological, and social systems. He related servomechanisms to the "behavioristic approach" to organisms.

In 1948 Wiener published his foremost achievement: *Cybernetics, or Control and Communication in the Animal and the Machine*. It is an interdisciplinary study connecting the fields of control systems, electrical network theory, and mechanical engineering with the fields of biology and neuroscience. In 1950 Wiener published his popular book: *The Human Use of Human Beings: Cybernetics and Society*. It examined the social implications of cybernetics, drawing analogies between automatic systems (as exemplified by the regulated steam engine) and human institutions.

Wiener fostered an evolving understanding that the boundary between humans and machines affected the performance of dynamic systems. He encouraged the sentiment that cybernetics was a fruitful area of research. Wiener's efforts to bring his model to broad communities of physiologists, physicians, and social scientists, are well documented. Through the informal "Teleological Society," the series of Macy Conferences, and a growing identity as a public intellectual, Wiener elevated his thinking on control and communication to a moral philosophy of technology, and enjoyed enthusiastic response.

Animal and the machine

Plato (429 BC –347 BC) is one of the most influential authors in western philosophy. Socrates appears in many of Plato's works. In Plato's book Alcibiades, SOCRATES asks: "In a ship, if a man having the power to do what he likes, has no intelligence or skill in steering (=*kybernetikos*), do you see what will happen to him and to his fellow-sailors?" ALCIBIADES answers:" Yes; I see that they will all perish." Plato's word *kybernetikos* (skill at steering) is from *kybernao* (I steer). This word, which is also spelled *kubernao*, is the stem for the Latin word *gubernator*. In turn, the Latin word *gubernator* through French became the English word *governor*.

One usage of the word *governor* is to denote a speed limiter; that is, a mechanism that governs the speed of an engine. In 1948 Professor Norbert Wiener published his book *Cybernetics: Control and Communication in the Animal and the Machine*. Central to his thinking were servomechanisms used in communication and control. Of course, the obvious title of the book would have been *Servomechanisms: Control and Communication in the Animal and the Machine*. The old term for servomechanism is the word "governor." As we have seen the word governor comes from Plato's term *kybernetikes*. In the classical period the Roman letter 'c' became the equivalent of the Greek kappa, so *kybernetikes* is also spelled as *cybernetices*. Wiener acknowledged his indebtedness to Maxwell's *governor*, but he chose to use Plato's more mysterious word *cybernetics* for the new discipline.

To the potential reader, the word *servomechanisms* in the title of Wiener's book would have had the effect of limiting the scope of the book to an engineering methodology. Wiener's intent was to emphasize that servomechanisms are used not only to control machinery but also to regulate animals. The connection between animal and machine is the essence of cybernetics. The skills of scientists and engineers are needed not only to construct manufactured servomechanisms used for machinery, but also to understand the natural servomechanisms used to attain homeostasis in animals. Wiener's book had immediate appeal and gained wide readership. Wiener imparted to the word *cybernetics* a mixture of elucidation and elegance that made its place secure in language and in science.

Cybernetics is a broad field. It includes communication, control and computing. It is also vital in remote sensing as is carried out by radar installations and other detecting devices. Cybernetics draws upon his fundamental work in the theory of filtering and prediction. Wiener presents an overall vision of the importance of a statistical approach to the overall physical and biological environment. A probabilistic current sweeps through Wiener's cybernetics to form broad theoretical and practical framework. Each of the many features that Wiener includes in cybernetics has important applications. Brought together, they present a comprehensive view of how things work.

Cybernetics provides a way making sense of the overall field of automation. In such a perspective, cybernetics deals with the essence of modern technology. It includes both man and machine. Experimental psychology or neurophysiology readily fit into the scheme of cybernetics. However, Wiener goes so far as to also include sociological, philosophical, and ethical problems as appropriate areas in the field of cybernetics. In ways this generality proved to be a disadvantage to getting cybernetics adopted as separate field of science. Cybernetics is so broad and all-embracing that few scientists list themselves as working in the field. Instead they use more specialized areas such as communication, automatic machines, etc. Even the narrowly technical content of the fields associated with the word cybernetics is extensive. Wiener goes on to include any phenomena of life which embodies feedback in behavioral and regulatory functions of the body. Wiener even goes further. He says that sociology, anthropology and economics are primarily sciences of communication and therefore fall under the general head of cybernetics.

A servomechanism is a device which acts continually on the basis of information to attain a specified goal in the face of A servomechanism makes use of feedback. Feedback comes in two varieties: (1) negative feedback and (2) positive feedback. Feedback is called negative when it decreases any departure from a given bearing. Feedback is called positive when it increases any departure from a given bearing. A positive feedback mechanism is the exact opposite of a negative feedback mechanism. With negative feedback, the output reduces the original stimulus. In a positive feedback system, the output enhances the original stimulus.

Negative feedback can be used to make the large output signal of an amplifier conform closely in shape to the small input. Negative feedback amplifiers were extremely important in communication systems long before the day of cybernetics. An example of negative feedback would be the bodily mechanism that controls of glucose (blood sugar) by insulin. When blood sugar rises, receptors in the body detect the change. As a result, the pancreas secretes insulin into the blood. The effect is the lowering of glucose. Once the glucose level reaches homeostasis, the pancreas stops releasing insulin.

The key is that a negative feedback mechanism inhibits the original stimulus and a positive feedback mechanism enhances it. An example of a positive feedback would be the mechanism of blood clotting. A vessel is damaged. Platelets begin clinging to the injured place. Chemicals are released that attract more platelets. The platelets keep accumulating and releasing chemicals until a clot is made. In his book Cybernetics, Wiener puts great emphasis on negative feedback as an element of nervous control and on its failure as an explanation of disabilities, such as tremors of the hand, which are ascribed to failures of a negative feedback system of the body.

We can easily think of other examples of feedback. The thermostat on the wall uses negative feedback. The autopilots of airplanes use negative feedback in manipulating the controls in order to keep the compass and altimeter readings at assigned values. The thermostat on the wall measures the temperature of the room and turns the furnace off or on so as to maintain the temperature at a constant value.

When we walk carrying a tray of water, we may be tempted to watch the water in the tray and try to tilt the tray so as to keep the water from spilling. This is often disastrous. The more we tilt the tray to avoid spilling the water, the more wildly the water may slosh about. When we apply feedback so as to change a process on the basis of its observed state, the over-all situation may be unstable. That is, instead of reducing small deviations from the desired goal, the control we exert may make them larger.

Negative feedback devices can be unstable; the effect of the output can sometimes be to make the behavior diverge widely from the desired goal. Wiener attributes tremors and some other malfunctioning of the human being to improper functioning of negative feedback mechanisms. The possibility of instability is a particularly hazardous matter in feedback circuits. The thing we do to make corrections most complete and perfect is to make the feedback stronger. But this is the very thing that tends to make the system unstable. Of course, an unstable system is not good. An unstable system can result in such behavior as an airplane or missile veering wildly instead of following the target, the temperature of a room rising and falling rapidly, an engine racing or coming to a stop, or an amplifier producing a singing output of high amplitude when there is no input.

Human beings use negative feedback in controling their motions to achieve certain ends. For example, a person uses negative feedback from eye to hand in guiding a pen across the paper, and negative feedback from ear to tongue and lips in learning to speak. The animal uses negative feedback to maintain bodily temperature despite the outside temperature. It uses negative feedback to maintain constant chemical properties of the blood and tissues. Homeostasis is the ability to keep body within a narrow range of limits despite environmental conditions outside of the body. There are many negative feedback pathways in biological systems, including: temperature regulation, blood pressure regulation, blood sugar regulation, and thyroid regulation.

The physiological system of a higher animal must maintain internal stability. Homeostasis refers to the ability of an organism or a cell to regulate its internal conditions so as to stabilize health and functioning, regardless of the outside conditions. Homeostasis is achieved by an arrangement of feedback loops. Homeostasis preserves the internal state at which the human body operates best. The human body has its own internal controllers for maintaining its temperature, pH, hormone levels, blood sugar and other internal variable levels. We sweat to cool off when hot, and we shiver to warm up when cold. In entomology, homeostasis is the ability of members of a colony of social insects to behave cooperatively to produce a desired result, such as when bees coordinate the fanning of their wings to cool the hive.

In its most fundamental sense, cybernetics deals with the problems that confront a steersman. A steersman is one that steers a ship. Sometime in the period of the sixth and seventh centuries, accounts of the voyages of the Celtic priest St. Brendon were composed. They give explicit descriptions of northern latitudes and demonstrate knowledge of navigation to Iceland and Greenland. In the early seventeenth century, Captain John Smith of Alford, Lincolnshire, explored the North American coast. His charts made possible the settlement of New England. In such cases, the skill and astuteness of the steersman represented the foundations of cybernetics.

The Blaise Pascal Medal for Science and Technology was awarded to Enders A. Robinson as "the father of digital geophysics." The Medal was awarded by the European Academy of Sciences and presented in Brussels, Belgium on October 22, 2004 at the Opening Ceremony of the Annual Congress on Nanotechnologies.

Chapter 8. Minimum delay

Shakespeare (*Twelfth Night):* What's to come is still unsure. In delay there lies no plenty.

Shakespeare's plays

The plot of a play implies opposition between forces. The struggle may be external or internal. Shakespeare's play *Macbeth* tells the story of a brave Scottish general named Macbeth. At the beginning of the play, Macbeth is bewitched by a trio of witches who prophesy that he will become King of Scotland. The bewitchment induces Macbeth into a series of crimes, each one bringing him further down into the depths of despair. He murders King Duncan and takes the throne for himself. He has his longtime companion Banquo killed. He has the wife and son of Macduff murdered. At the end of the play Macbeth realizes that he was bewitched. He exclaims:

And be these juggling fiends [the witches] no more believed,
That palter with us in a double sense;
That keep the word of promise to our ear,
And break it to our hope.

Macbeth with all hope gone is slain by Macduff. The throne is returned to the son of King Duncan. Macbeth as victorious general is at his highest point at the beginning of the play. From there he continuously descends with no procrastination. For that reason, the tragedy of Macbeth is minimum delay.

The tragedy of Hamlet is not minimum delay. Hamlet is a procrastinator. That characteristic leads Hamlet into a series of delays, which elongate the time before he avenges his murdered father. The denouement comes in the final scene when Hamlet dies:

Now cracks a noble heart. Good night sweet prince:
And flights of angels sing thee to thy rest

Shakespeare's comedy *As You Like* It is maximum delay. All of the good results are delayed until the final scene in the forest. In this scene couples marry, an evildoer repents, two disguised characters are revealed, and the good Duke is restored to power.

In Shakespeare's play *King Henry 6, Part 1*, the Duke of York and the Duke of Somerset are English lords engaged in a war against France. The play takes place at the beginning of the English War of the Roses, which was a dynastic war for the throne of England. The war was fought between the House of Lancaster (red rose) and the house of York (white rose). The Duke of York needs more cavalry in order to win the upcoming battle against the French. York asks Somerset, who is on the Lancaster side, to supply the needed cavalry. Somerset does not say no, but he delays. In other words, Somerset uses Negation by Delay. It does work. The French win the battle and thereby regain French land (Maine, Blois, Poitiers, and Tours) occupied by the English. The Duke of York laments:

> Maine, Blois, Poitiers, and Tours are won away,
> 'Long all of Somerset and his delay.

The lesson is this. From the nature of things, some minimum amount of delay is unavoidable. However, any extra delay (procrastination) in addition to this minimum delay can be harmless in some cases; detrimental on other cases.

Parkinson 's "Law of Delay"

Cyril Northcote Parkinson (1909-1993) was a British naval historian and author of many books. Parkinson's best known law states that "work expands so as to fill the time available for its completion." Parkinson also wrote that there is nothing static in our changing world. In order to obtain a favorable result, a person or system should give a positive ("Yes") answer with minimum delay. A "No" answer would be harmful. Parkinson points out that the *Prohibitive Procrastinator* never says "No." Instead he says, "In due course." This expression foreshadows *Negation by Delay*. In other words, the Prohibitive Procrastinator does not act with minimum delay. Parkinson's "Law of Delay" states that the introduction of extra delay can be enough to produce negation. Parkinson's "Law of Delay" is essential to the understanding of feedback.

The use of negation by delay is based upon the calculation of what amount of delay will equal negation. A man needs money to buy an item available only today. You reply "I will lend the amount to you in due course." You make the money available the next day. However, by then the man no longer needs the money for the item is gone. Such procrastination is equivalent to "no," because the delay is greater than the period for which the item is available.

Let us give another example. Take the unit of time as one calendar day. Suppose you deposit $1000 in The First National Bank on Monday. Later, on that same day, you need $1000 at once to close a deal. The First National Bank lets you withdraw the deposit that you just made, and you close the deal.

Let us continue along these lines. Suppose you deposit $1000 in The Second National Bank on Monday. Later, on that same day, you need the money at once for a deal. The deal must be closed on Monday. The Second National Bank says: "In due course! You must wait one day before you can withdraw any money deposited. You cannot withdraw your deposit of $1000 until Tuesday." In such a case, the bank acts with an extra delay of one day. That extra day would make you lose the deal.

Let us look at a hypothetical example from Wall Street. Say it is 9:30 AM on a gorgeous summer morning in New York. An executive presses the button at the New York Stock Exchange that signals the opening the trading day. A millisecond is one thousandth of a second. A few milliseconds after the opening, buy and sell orders are coming across the market's servers with alarming speed. These trades are obviously unusual. They come in small batches of 100 shares that involve nearly 150 different financial products. Computer programs deployed by financial firms swoop in. They buy undervalued stocks as the unusual sales drive their prices down. They sell overvalued stocks as the purchases drive their prices up. The algorithms are making a killing.

What is happening? We must look to a certain financial company across the river in New Jersey. This company is in absolute panic. By mistake the company's trading program has gone rogue. The company futilely tries to deactivate it. However, the program keeps making detrimental trades orders that are costing the company $10 million per minute. No one at the company knows how to shut the program down. Finally the computer guru wanders in to work still half asleep. Within seconds the guru shuts the runaway program down. The fiasco had been going on for 50 minutes. This 50 minute delay cost the company half a billion dollars. The lesson learned was that the company must be able to shut down any trading program with minimum delay.

New technologies have changed Wall Street. Today's markets are wilder, less transparent, and, most importantly, faster than ever before. Stock exchanges can now execute trades in less than a half a millionth of a second. This is more than a million times faster than the human mind can make a decision. An algorithm might make only a tiny fraction of a cent on each trade. However, by making a great many trades, the profits can add up. We can safely say that trading on the New York Stock Exchange is about as close to the ideal of minimum delay as you can get.

Impulse response

The impulse response of a system is the output of the system resulting from a brief input signal, called an impulse. Other words for impulse are jolt, spike, blow, shock, innovation. The shock treatment is used all the time by animals. Before you dive into the water, you stick your toe in. If cold shivers run up and down your spine, you do not dive into the water. The cold from the quick dipping of the toe represents the impulse. The piercing cold shivers represent the impulse response.

A thermostat is based upon feedback. The room temperature is fed back and compared to setpoint (desired indoor temperature). The deviation is the difference between room temperature and setpoint. The deviation is kept small by using the deviation to control the fuel supply to the furnace. Suppose you want to test the effectiveness of a heating system. Let us describe the impulse method. First you subject the system to an impulse. You may do so as follows. Open some doors and windows and let the heat escape. As soon as the temperature falls to some specific temperature (e.g., ten degrees below setpoint), close the doors and windows. This jolt in room temperature represents an impulse. How does the thermostat react to such an impulse? The reaction is called the impulse response. Say that we look at the room temperature every five minutes, which we call a unit of time.

At time 0 the deviation is 10 degrees below setpoint.
At time 1 suppose the deviation is 5 degrees below setpoint.
At time 2, suppose the deviation is 2.5 degrees below setpoint.
At time 3, suppose the deviation is 1.25 degrees below setpoint.
At time 4, suppose the deviation is 0.625 degrees below setpoint.

Because this last reading is so close to the setpoint, we can say that he setpoint has been reached for all practical purposes. In summary, after the jolt, it takes about 20 minutes for the heater to bring the room temperature up to setpoint. If we collect all the numbers, we obtain the impulse response; namely, the series (with members takes at five-minute intervals)

1, 5, 2.5, 1.25, 0.625

The impulse response shows that the room temperature does not instantaneously recover from the jolt. Specifically, for a jolt of ten degrees, it takes 20 minutes of heating for the room temperature to rise again to setpoint. In other words, it takes time to get the job done. This impulse response is a geometrically decreasing. In continuous time it would be called an exponentially decreasing. A geometrically decreasing series is the textbook example of a minimum delay impulse response. (In electrical engineering, the term *minimum phase* is used instead of the term *minimum delay*.)

Minimum delay and extra delay

Extra delay comes from putting off a decision (a.k.a. procrastinating). Let us give an example of minimum delay and extra delay. We have a hot water heater. The water in the tank is kept at a temperature of 110 degrees. Attached to the tank is a faucet. Some water is held within the faucet. Since the faucet has not been used for some time, the temperature of the water inside the faucet itself is only 70 degrees. You turn the faucet on. It takes 3 seconds to expel the cold water. After that delay, the water comes out at 110 degrees. The 3 second delay is the minimum delay.

What is extra delay (a.k.a. procrastination)? From the tank, suppose there is a pipe to the kitchen faucet. The water temperature in the pipe and kitchen faucet is 70 degrees. You turn the kitchen faucet on. Suppose it takes 9 seconds to expel the cold water in the pipe plus 3 more seconds for the kitchen faucet. After a delay of 9 plus 3, or 12 seconds, the water comes out at 110 degrees. Let us now analyze the situation. The faucet on the water tank and the kitchen faucet are both perfectly stable systems that give the same hot water. However, the kitchen faucet has extra delay of 9 seconds. In this case the extra delay is inconvenient but it is not critical.

When driving an automobile, a person notes the discrepancy, or error, between his steering and the direction of the road. If the error is to the left, he turns the steering wheel to the right, and vice versa. There is always a time-delay between the instant the human operator observes an error and the instant he starts to take corrective action. The operator with the minimum-delay is the safest driver.

There are examples in which extra delay is dangerous. A driver of a car sees a fallen tree on the road. He already has both hands on the wheel and he quickly applies the brakes. The delay represents minimum delay. However, suppose the driver is engaged in a discussion on the cell phone. As a result there is an extra delay required to put aside the cell phone and apply the brakes. This extra delay might result in an accident.

Let us take another example. On an airplane, the flight attendant gives instructions. In case of emergency, oxygen masks will drop down. The passenger should always fit his or her own mask before helping children, the disabled, or persons requiring assistance. It is most important that the passenger gets his mask on with as little delay as possible. Otherwise he or she might become unconscious and then could not help anyone else. The instruction given by the flight attendant is that the passenger should act with minimum delay. However, suppose the passenger does not heed the instructions. The passenger puts on the child's mask first, and then put his or her mask on. That would be a non-minimum delay response. It involves the extra delay in putting on the child's mask. In some cases the passenger can get away with this extra delay. In other case it could be fatal.

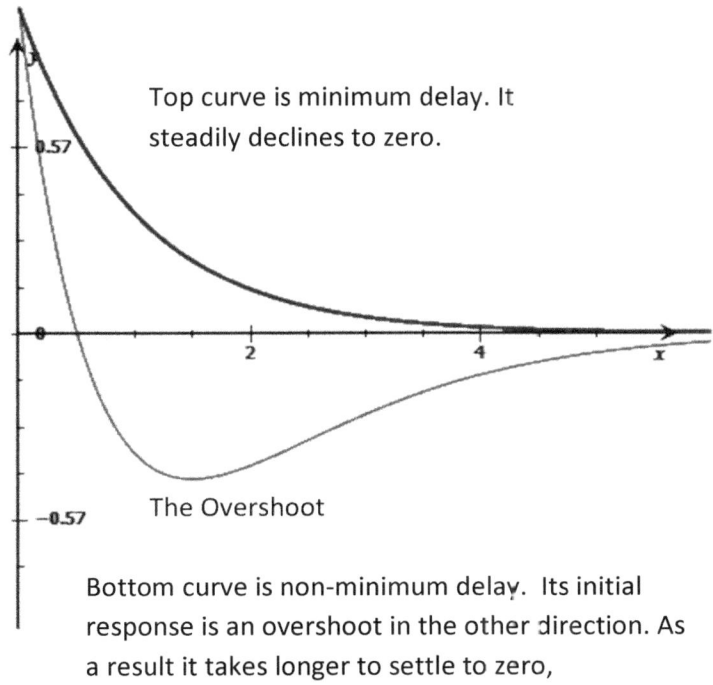

Top curve is minimum delay. It steadily declines to zero.

The Overshoot

Bottom curve is non-minimum delay. Its initial response is an overshoot in the other direction. As a result it takes longer to settle to zero,

Figure 6. Top curve is minimum delay. Bottom curve is non-minimum delay.

A self-driving car is constructed. One test involves the car performing a turn onto a cross street. A feedback mechanism indicates the deviation between the desired direction and the actual direction of the car. This error activates the guidance system consisting of power amplifiers which force the steering wheel in the direction that decreases the deviation. It takes a period of time for the car to make a smooth turn to the correct direction. We shall examine two cases of what could happen.

Case 1. Gentle turn. See top curve in Figure 6. The feedback mechanism steers the wheel so the car smoothly turns onto the side street.

Case 2. Abrupt turn. See bottom curve in Figure 6. The feedback mechanism turns the car so abruptly that the car overshoots the cross street. Now the feedback mechanism must turn the car in the opposite direction so that the car comes back onto the side street.

The first case is minimum-delay. The car is eased into the turn so that it steadily get on the cross street without overshooting. The second case is not minimum-delay. There is extra delay in addition to minimum-delay. The car is turned too powerfully at first so that it overshoots the cross street. Now the car must be turned the other way to get onto the cross street.

As we have just seen a non-minimum-delay system can overshoot its mark. In other words its initial behavior could be a swing to the opposite direction. The system then has to correct itself. We can take bacteria as an example. When bacteria have access to more food, they devote themselves to eating and fail to reproduce themselves. Because of natural deaths, the number of bacteria at first decreases. As the bacteria get stronger from eating, they start reproducing and their numbers increase to greater numbers than before. In summary, with more food, the number of bacteria initially decreases, but then increases.

The hiring of a new employee could decrease productivity at first due to the training required. However in the long run productivity should increase. In other words, short term losses can lead to long-term gains. You have to spend money to make money.

Consider the heating of a house with a wood-burning fireplace. If the temperature is too low, you add more wood to build the fire up. However, at first you achieve the opposite result. The temperature is reduced, because the added logs damp the fire. As these logs start to burn, the fire gets larger and the temperature begins to rise.

In general, digestion is non-minimum delay in terms of energy. In order to gain calories from food, an animal must expend calories to break the food down. In other words, the initial output is in the wrong direction. However, in due time the animal gains more energy from the food than was expended.

Many medical treatments might seem to be non-minimum delay. For example, cancer chemotherapy requires toxic poisons. Let us consider how the patient feels. Initially the chemotherapy makes the patient feel worse, but the patient feels much better in the long run with the cancer gone. In terms of the patient's feelings, chemotherapy is non-minimum delay. Let us consider the cancer cells. The chemotherapy never lets the cancer cells increase. Instead it steadily kills the cancer cells from the outset. In terms of killing cancer cells, the chemotherapy is minimum delay. In engineer practice, the term "minimum-phase" is used in place of the term "minimum delay." Both terms mean the same thing.

Chapter 9. Feedforward and feedback

> Gottfried Leibniz: I maintain also that substances, whether
> material or immaterial, cannot be conceived in their bare
> essence without any activity, activity being of the essence of
> substance in general.

Feedforward (past inputs) vs. feedback (past outputs)

A digital signal is a sequence of numerical values whose variations depend upon time coordinates and/or space coordinates. A system is like black box. On the left side is the input channel and on the right side is the output channel. Time is measured in discrete equally-spaced units; for example, in seconds. For any given present time, there are past times and future times. We can know present and past values of a signal but not future values.

The simplest possible system is the one which has a direct passage from input to output. For such a system the present output is depends upon the present input. Such a system requires no memory of the past. A (purely) feedforward system remembers only past input values. A (purely) feedback system remembers only past output values. A (mixed) feedforward -feedback system remembers both past input values and past output values. The power of the digital computer comes from its memory. Everything stored in memory can be used to influence present action. Natural analog systems (such as the brain) have memory, but so far humans have little or no understanding of how they work.

We assume that all systems have a direct channel from input to output. Thus the output will always depend upon the present input. However the system has the ability to modify its output by means of its memory. That is why memory is so important. A feedforward system can modify its output by past input values. A feedback system can modify its output by past output values. A feedforward-feedback system can modify its output by both.

Once a gun is shot, the shell travels on a ballistic trajectory. Nothing can be done on the ground to change the course of the shell. In contrast a cruise missile trajectory is adjusted according to its path of travel by a feedback mechanism. Dead reckoning such as done by an inertial navigation system (INS) is a good example of feedforward. Positional reckoning such as done by a global positioning system (GPS) is a good example of feedback.

A **feedforward control system** requires that **past inputs** be retained in memory. The system uses the present input together with these memorized past inputs to determine the present output. The system never takes past outputs into account. A person gives money each year to a charity because he wants to support the charity as much as he can. The charity always needs money. The person uses his list of past contributions to help him determine what amount to give this year. His behavior is a feedforward process. His decision as to what amount to donate is solely dependent on what he can donate in the present and on what he has donated in the past. The output is the good service rendered by the charity. The person does not keep track of any particular service. Past outputs have no direct effect on the present input (the person's donation).

A **feedback control system** requires that **past outputs** be retained in memory. The system uses the present input together with these memorized past outputs to determine the present output. The system never takes past inputs into account. The system is error-based. It uses past outputs to provide a means of reducing the error between the setpoint and the output. A person is driving on a twisty road. The input is his movement of the steering wheel. The output is the direction of the car on the road. He does remember any of the past movements that he has imparted to the steering wheel. He is only concerned with the present movement. The present movement will be determined according to how well the direction of the car (past outputs) followed the direction road (setpoint). Past inputs have no direct effect on the present input (movement of the steering wheel).

Positive and negative feedback

Feedback control is found in many situations. Control is exercised by negative feedback, which is to say that the information fed back is the amount of departure from the desired condition. Negative feedback occurs when the result of a process influences the operation of the process itself in such a way as to reduce changes. Negative feedback tends to make a system self-regulating; it can produce stability and reduce the effect of fluctuations. In systems controlled by a negative feedback loop, a measurement of some variable is compared with a required value to estimate the current error, which is then used to reduce the gap between the measurement and the required value, as well as to counter the effects of unpredictable influences from the system's environment.

Positive feedback is a process in which the effects of a small disturbance on a system include an increase in the magnitude of the perturbation. In contrast, a system in which the results of a change act to reduce or counteract it has negative feedback.

The terms "positive" and "negative" were first applied to feedback prior to World War 2. A positive interest rate, say 8 percent per annum, increases the value of a bank account. For example an initial deposit of 1 dollar would become 1.08 after one year, 1.21 after two years, 1.331 after three years.   Paying interest means a positive rate. A positive rate means positive feedback. Left in the bank, the balance would eventually exceed any amount that the bank could ever amass. In such a case the bank could not meet its obligation to liquidate the bank account. Positive feedback is unstable from the bank's point of view. Banks prefer to pay the smallest interest rate possible.

A fee, say 8 percent per annum, decreases the value of an account. For example an initial deposit of 1 dollar would become 0.92 after one year, 0.8464 after two years, 0.778 after three years. Charging a fee means a negative rate. A negative rate means negative feedback. Left in the bank, the balance would eventually be less than any amount. Negative feedback is stable from the bank's point of view. Banks prefer to charge the highest fee rate possible.

In 1625, the Dutch West India Company bought Manhattan from the Canarsees for beads and trinkets worth about $16. Since Manhattan real estate is expensive today, it would seem that the Canarsees made a unfortunate sale. However, suppose the Canarsees had sold the beads and trinkets for $16 and deposited that amount with the Dutch West India Company at the rate of 8 percent interest compounded annually. (Paying interest means a positive rate).Their balance would grow over time, and now it would be enough to buy back all of Manhattan with money left over. That is the power of positive feedback. Because the output grows without limit, positive feedback is unstable. The Dutch West India Company would become insolvent long before it could pay back such an amount.

In those bygone days, there was always a fear of pirates and advancing English armies. Suppose that the Dutch West India Company offered the Canarsees a deal to hold the $16 safely against such threats. For this service the Dutch West India Company would charge a fee at the rate of 8 percent compounded annually. (Charging a fee means a negative rate). The money that the Canarsees would have now would not be enough to buy anything at all. That is the power of negative feedback. The output decreases without limit. For this reason, negative feedback is stable. The Dutch West India Company would remain solvent after it paid such an infinitesimal amount.

In the nineteenth century, James Clerk Maxwell described several kinds of "component motions" associated with the centrifugal governors used in steam engines, distinguishing between those that lead to a continual increase in a disturbance or the amplitude of an oscillation (positive feedback), and those that lead to a decrease of the same (negative feedback). The terms positive feedback and negative feedback are not based on the sign of the feedback itself but rather on the rate. Positive feedback has a positive rate. Negative feedback has a negative rate. On each line in the table below, there are various terms for dual-designation. They are all different ways of saying the same thing.

positive feedback	negative feedback,
unstable feedback	stable feedback
gap-enhancing feedback	gap-reducing feedback,
self-reinforcing feedback	self-correcting feedback,
reinforcing feedback	balancing feedback,
regenerative feedback	degenerative feedback

As we have seen, the gap (or deviation) is the difference between actual value and setpoint. In driving a car, if the gap goes down (in magnitude), say from −10 to 5 to 0, then the feedback is gap-reducing (or stable or negative). On the other had if the gap goes up (in magnitude), say from 5 to −20 to 40, then the feedback is gap-enhancing (or unstable or positive).

It is important to identify configurations that utilize feedback. In the industrial revolution, steam power came into its own. The engine-governor and other automatic mechanisms were used to regulate steam power. Of particular importance was the steering servo-engine on ships, which operated the rudder in correspondence with movements of the helm. These devices, and others such as simple voltage regulators, make use of feedback. The problems arising from new technology have required new kinds of feedback procedures. Automatic regulators in chemical plants control temperature and flow.   Feedback circuits give accurate amplification of radio signals. The study of automatic control systems yielded new insight into many systems in nature and in humans. Feedback explains how the heart beats, how an animal stands upright without falling, and how an animal population regularly fluctuates between scarcity and abundance.

Negative feedback from eye to hand is used in guiding a pen across the paper. Negative feedback is used from ear to tongue in learning to speak. The animal organism makes use of negative feedback in many other ways. Negative feedback is used to maintain body temperature despite changes in environmental temperature. Negative feedback maintains constant chemical properties of the blood and tissues. The ability of the body to maintain a narrow range of conditions despite environmental changes is called *homeostasis*. In his book *Cybernetics*, Wiener puts great emphasis on negative feedback as an element of nervous control. Disabilities, such as tremors of the hand, are ascribed to failures of negative feedback systems in the body. Human beings learn by doing and by thinking about what they have done.

A major part of cybernetics is negative feedback. A thermostat makes use of negative feedback when it measures the temperature of a room and starts or stops the furnace in order to make the temperature conform to a specified value. The autopilots of airplanes use negative feedback in manipulating the controls in order to keep the compass and altimeter readings at assigned values. Human beings use negative feedback in controlling the motions of their hands to achieve certain ends. Negative feedback can also be used in order to make the large output signal of an amplifier conform closely in shape to the small input. Negative feedback amplifiers were extremely important in communication systems long before the day of cybernetics.

Finally, cybernetics has laid claim to the whole field of automata complex machines, including telephone switching systems, which have been in existence for many years, and electronic computers, which have been with us only since World War 2. If all this is so, cybernetics includes most of the essence of modern technology, excluding the brute production and use of power. It includes our knowledge of the organization and function man as well. Cybernetics almost becomes another word for all of most intriguing problems of the world. As we have seen, Wiener includes sociological, philosophical, and ethical problems among these.

Feedforward in multistage rocket

A multistage rocket uses two or more stages, each of which contains its own engines and propellant. It is essentially two or more rockets stacked on top of or attached next to each other. Two-stage rockets are quite common, but rockets with as many as five separate stages have been successfully launched. By jettisoning stages when they run out of propellant, the mass of the remaining rocket is decreased. This allows the thrust of the remaining stages to accelerate the rocket more easily to its final speed and height. At time zero the first stage ignites sends the rocket on its way. At time one the second stage ignites giving the rocket an additional boost. The astronauts in the rocket as well as people on the ground have no way of steering the rocket during these propulsion stages. There is no feedback to alter the direction of the rocket if it veers off course. A multistage rocket is a feedforward system.

When one drives a car the propulsion is entirely different. The driver sets a desired speed in his mind. This speed is called the setpoint, say 60 mph. If the speedometer registers 50 mph, the driver presses down on the gas pedal to speed up. If he presses too much, the car could accelerate to 65 mph. In such a case the driver would have to let up on the gas pedal to decelerate down to the setpoint of 60 mph. The process of driving a car is a feedback system. The gap is the difference given by speed minus setpoint. In the example just given, the gap starts at 10 mph below setpoint. The driver pushes the accelerator down. The gap changes 5 mph above setpoint. The driver lets the accelerator up. Finally the gap reduces to zero.

Feedback occurs when outputs of a system are routed back as inputs as part of a chain of cause-and-effect that forms a circuit or loop. The system can then be said to feed back into itself. Feedback generally is the use of the gap between actual level and reference to make the gap smaller in magnitude.

Chapter 10. Direct and inverse

Gottfried Leibniz: When a truth is necessary, the reason for it can be found by analysis, that is, by resolving it into simpler ideas and truths until the primary ones are reached.

Direct and inverse

In life we are continually confronted with inverse problems. There is an automobile accident (the effect). The police seek the cause. Was it distracted driver, driver fatigue, drunk driving, speeding, aggressive driving, weather, or something else? Ancient people would see an eclipse of the moon (the effect). What was the cause? They believed it was an omen of the gods. It took eons before the cause was determined. A lunar eclipse occurs when the moon passes directly behind the Earth into its shadow.

In pharmacology, an agonist (i.e., the cause) is a molecule that binds to a receptor and activates the receptor to produce a biological response (i.e., the effect). A partial agonist has a weaker preference than an agonist for the same receptor and shifts the equilibrium to a smaller extent than an agonist. An antagonist reduces the effect of an agonist by preventing the agonist from binding to receptors. Whereas an agonist causes an action, an inverse agonist causes an action opposite to that of the agonist. In other words, an inverse agonist is an agent that binds to the same receptor an agonist but induces a pharmacological response opposite to the response of the agonist. An example is the cannabinoid inverse agonist named rimonabant.

Let us summarize. An inverse agonist is a molecule that binds to the same site as an agonist. However, it elicits the opposite effect to that of the agonist. In other words, an inverse agonist demonstrates negative efficacy. In conventional two stage drug receptor interaction, receptors can be in either the resting state or the active state. The binding of a molecule to the receptor causes a conformational change in the receptor. If it initiates activity, then it is an agonist with varying degrees of positive intrinsic activity. If it prevents the activity of an agonist, then it is an antagonist with zero intrinsic activity. If it has negative efficacy or negative intrinsic activity, then it is an inverse agonist.

Both antagonists and inverse agonists reduce the activity if a receptor and, in the presence of an agonist, reduce its activity. However, unlike inverse agonists, antagonists do not have any effect in the absence of an agonist. Inverse agonism was first discovered when B-carbolines and other compounds that were shown not only to have an antagonistic effect at benzodiazepine receptors but also to elicit an effect that was the opposite of the benzodiazepine in the absence of benzodiazepines. Currently there are a number of well-established true inverse agonists including antipsychotics, antidepressants and other psychopharmacological drugs that have inverse agonist activity at serotonin, dopamine, histamine, opioid, cannabinoid and muscarinic receptors.

MIT Professor Dirk J. Struik

A high point in undergraduate years at MIT was the course M442 in differential geometry during the fall-winter term of 1949-1950. It was given by Professor Dirk Struik. The text was a thick set of mimeographed notes which were to become his book *Lectures on Classical Differential Geometry* published in 1950. Struik writes in the Preface, "The author owes much to the constructive criticism of his class in M442 during the fall winter-term of 1949-1950, which is responsible for many an improvement in text and in problems."

Professor Dirk J. Struik was an eminent analyst and geometer and an internationally acclaimed historian of mathematics. He celebrated his 106th birthday on September 30, 2000. He entered the University of Leiden in 1912, and took courses in mathematics and physics from Hendrik Lorentz and Willem de Sitter. Professor Struik collaborated with MIT Professor Norbert Wiener, who brought him to MIT in 1926. From then until his retirement in 1960, Professor Struik was a member of the MIT mathematics faculty. In 1948 he published his distinguished book *A Concise History of Mathematics*. In 1972 he was made an honorary research associate in the History of Science Department at Harvard University.

In class Professor Struik once told us about Isaac Barrow (1630-1677). Barrow's works were fundamental to the establishment of a sound basis for the teaching of university mathematics in England. Barrow introduced the concept of the differential triangle, which in the limit gives the tangent to a curve. In Chapter 4 of his book *A Source Book in Mathematics*, Struik writes "We find the fundamental theorem of calculus in the *Lectiones Geometricae* by Isaac Barrow. His most famous student was Isaac Newton who succeeded him in the Chair." In Chapter 5 of his book, Struik writes "The fundamental step (in the final discovery of calculus) was taken by Isaac Newton and Gottfreid Wilhelm Leibniz."

It was Newton and Leibnitz who established calculus. However, Barrow proved the fundamental theorem of calculus before the discovery of calculus. This statement seems like a contradiction in terms. Let us explain. Barrow proved the fundamental theorem of calculus with Euclidean geometry, not with symbols. By inventing calculus, Newton and Leibnitz invented symbols that allowed people to do infinitesimal mathematics without the use of geometric pictures. Calculus (and more generally, the more comprehensive subject known as infinitesimal analysis) allows all kinds of mathematics to be down with symbols. In particular, geometry can be done with no pictures, but with symbols alone. For example, in William Graustein's *Differential Geometry* (1935), there are only 41 quite simple pictures in its 230 pages. The book is full of equations, one after another. The 41 pictures are not needed at all as far a mathematical rigor is concerned. Higher dimensional geometries are developed by means of equations not pictures.

Let us look at the superstring theory of modern physics. Instead of three spatial dimensions and one time dimension, the form of superstring theory known as M-theory requires ten spatial dimensions and one time dimension. In such a universe, reality is a cosmic substrate composed of eleven space-time dimensions. We humans cannot see these extra dimensions. Superstring theory says that we only see a meager slice of reality. It is easy to write the equation for an object in eleven dimensions; it would be essentially impossible to draw its geometric picture. Shakespeare wrote, "We are such stuff as dreams are made on." Modern physics says, "We are such stuff as mathematical equations are made on."

The pendulum swings back and forth. The digital computer allows us to solve equations that we could never solve before. However, it also allows us to depict pictures that we could never achieve before. Through such pictures, physicists, biologists, astronomers, and economists have created a new way of understanding the growth of complexity in nature. The new disciplines of chaos and fractals give a way of seeing order and pattern in what before seemed to be random and erratic. In the area of remote sensing, it is now possible to see intricate three dimensional computer-generated images of the interior of the human body. No longer do we need to read Dante to learn of the dreadful layers of underworld. Instead we can look at intricate images of the subterranean earth on a computer.

Ebb and flow

The moon controls the tides. On the ebb, water goes out and on the flow the water come back in. Flow is the inverse of ebb. Ebb is the inverse of flow. In his poem *Dover Beach*, Mathew Arnold (1822-1888) writes about the ebb and flow of the tide:

> Listen! you hear the grating roar
> Of pebbles which the waves draw back, and fling,
> At their return, up the high strand,
> Begin, and cease, and then again begin,
> With tremulous cadence slow, and bring
> The eternal note of sadness in.
> Sophocles long ago
> Heard it on the Aegean, and it brought
> Into his mind the turbid **ebb and flow**
> Of human misery.

In *The Unfading Beauty*, Thomas Carew speaks of a time for flames to waste away and a time to kindle never-dying fires.

> He that loves a rosy cheek,
>   Or a coral lip admires,
> Or from star-like eyes doth seek
>   Fuel to maintain his fires:
> As old Time makes these decay,
> So his **flames must waste away**.
>
> But a smooth and steadfast mind,
>   Gentle thoughts and calm desires,
> Hearts with equal love combined,
>   Kindle **never-dying fires**.
> Where these are not, I despise
> Lovely cheeks or lips or eye

In *L'Allegro*, John Milton speaks of a time for Mirth:

> HENCE, **loathèd Melancholy**,
> Of Cerberus and blackest Midnight born,
> But come, thou Goddess fair and free,
> In heaven yclep'd Euphrosyne,
> And by men, **heart-easing Mirth**.

In *Il Penseroso*, John Milton speaks of a time for Melancholy:

> HENCE, **vain deluding Joys**,
> The brood of Folly without father bred!
> But, hail! thou Goddess sage and holy!
> Hail, **divinest Melancholy**!

Heart-easing Mirth is the inverse of loathèd Melancholy. Divinest Melancholy is the inverse of vain deluding Joys.

The fundamental theorem

The fundamental theorem of calculus states that integration is the inverse of differentiation, and in turn differentiation is the inverse of integration. It is the mathematical version of *Ecclesiastes 3*:

> To every thing there is a season, and a time to every purpose under the heaven. A time to break down, and a time to build up.

"To break down" (i.e., differentiate) and "to build up" (i.e., integrate) are inverses of each other. In effect, **one cancels the other**. "Break down" does away with "build up," and vice versa. **If you want each one to cancel the other one without delay, then both must be minimum-delay**. In familiar terms, there is a time to spend, and a time to save. If you start out with a given amount, and from then on you spend as fast as you save, and save as fast as you spend, you will always retain the same amount. This situation represents minimum-delay spending and minimum-delay saving. On the other hand if you wait a few days before you spend, there will be extra delay (i.e., procrastination). The amount you have will vary from day to day. This situation represents a non-minimum-delay case.

The fundamental theorem turns from one thing to the inverse thing. To everything, turn, turn, turn. There is a thing and there is the inverse thing. Analysis is the inverse of synthesis. Subtraction is the inverse of addition. Division is the inverse of multiplication. Differentiation is the inverse of integration. Deconvolution is the inverse of convolution. **If you want to get back to where you started without delay, the convolution and the deconvolution must each be minimum delay.** Impulses from the various reflecting interfaces occur in a given time sequence. Passage though the earth convolves the Impulses with the reverberation wavelet. The reverberation wavelet is minimum delay. The deconvolution operator, which is the inverse of the reverberation wavelet, is also minimum delay. As a result deconvolution gets us back without delay to where things started; namely, it gets us back to the Impulses from the various reflecting interfaces.

Chapter 11. Prediction and filtering

Archimedes: Man has always learned from the past. After all, you cannot learn history in reverse!

MIT Professor George Valley

During World War 2 MIT changed dramatically from engineering instruction and scientific research to that of advancing the techniques of modern warfare. A great variety of projects for government agencies and industries were undertaken. Radar, servomechanisms, computing devices, guided missiles, special fuels, and metallurgical research, turbines and guns were among the more important projects. MIT also participated in the vast specialized training programs for the Army and Navy officers. I entered MIT as a first-year student in September 1946 when MIT resumed its regular university schedule. MIT admitted a great number of veterans, including any veteran who had taken any wartime class at MIT and who wanted to return to obtain a degree. MIT's energy now became concentrated on the future of technology.

It was an honor to be in a recitation class in physics taught Professor George E. Valley in 1946-1947. The lectures had a large number of students; the recitation classes had a small number. Prof. Valley knew how to teach students to solve the difficult problems assigned as homework. In fact, it seemed that MIT, for the most part, was interested in problem-solving and experimental design. There was not very much discussion about descriptive physics; mathematics took the place of words. The following excerpts are taken from an obituary appearing in the *MIT News* of October 20, 1999.

Professor George Edward Valley received his PhD in nuclear physics in 1939. He made the first measurements of iron isotopic abundance in terrestrial and meteoritic materials from 1939-1941 as a research associate and National Research Council Fellow at Harvard University. At MIT's Radiation Laboratory (1940-1946), he developed the H2X radar bombsight that played a strategic role in World War 2, permitting all-weather bombing. This required trips from Boston to London by bomber (passengers were required to ride in the empty bomb bay during takeoff) and return by seaplane via Brazil. In 1946, Dr. Valley became a professor of physics at MIT, where he conducted research in cosmic ray physics that included operation of a high-pressure cloud chamber at 12,000 feet on Mt. Evans, Colorado.

His interests continued in air defense as a member of the US Air Force Scientific Advisory Board (1946-64). **He conceived a national air defense system of radar stations guarding the northern air approaches to the United States and linked by telephone lines to a centralized digital computer (SAGE).** This work led directly to the founding of Lincoln Laboratory, where he was assistant and associate director (1949-1957), and to many advances in computer technology, including the invention of magnetic core memory and telephone transmission of digital data. In 1969, Professor Valley founded the Experimental Study Group, which emphasized a revolutionary learning process for first year students. "I was interested in fostering students who would be courageous enough to think in new ways without fear of getting the wrong answer," he later recalled.

In 1949 and 1950 Professor Valley went to great effort to save the unfinished MIT Whirlwind digital computer from an early demise. He was responsible for making Whirlwind the centralized digital computer for a continent-wide radar network. In 1951 Whirlwind was made available to MIT faculty and students. At that time MIT was unique in that no other university had a digital computer available to students. The Geophysical Analysis Group was among the first to take advantage of this great resource.

MIT Professor George Wadsworth

Wiener's report, "Extrapolation, Interpolation, and Smoothing of Stationary Time Series" was published for restricted circulation in

February 1942. It was largely unappreciated because it could not be implemented with existing analog machinery. However, Wiener's work did not go unappreciated in the Mathematics Department. Professor George Wadsworth was ready to make use of Wiener's prediction methods. Wadsworth would use digital methods and solve the equations on hand-operated desk calculators. Weather forecasting was not a split-second real-time operation as was the case for antiaircraft artillery. Instead the forecasts could be made in a matter of days or weeks during which time computations could be made.

In June 1942 the Army Air Forces Weather Division negotiated a contract (N30-053-AC-1065) with MIT to perform statistical analyses of meteorological and climatological data, particularly in relationship to weather forecasting. This research effort extended over a period of about four years. Professor Wadsworth was in charge of the project. Joseph G. Bryan and Chester H. Gordon were the two MIT colleagues who were most closely associated with this work during the entire period. The three continually kept Prof. Wiener informed of their progress.

Wadsworth, Bryan and Gordon computed many cases of the prediction of a single time-series and the generalization of the technique to the joint prediction of several time-series. An example of a single time-series is the sequence of pressure observations taken every 24 hours at a single station. Wadsworth, Bryan and Gordon assumed an additive model, namely the observed message is the addition of (1) a sequence that has certain inherent dynamic properties plus (2) a superimposed random component. This random component may be due to other variables not considered. The statistical problem is to separate the dynamics from the random element. Wadsworth and his colleagues wrote that this situation is entirely analogous to a similar one encountered when information is transmitted by mechanical or electrical methods. In this case the signal frequently becomes distorted, and when this distortion phenomenon is of a purely random statistical nature it is called "noise". Hence, the message contains the original signal plus a random noise. The problem is to isolate the signal from the message.

Wadsworth, Bryan and Gordon made a rather critical examination of Wiener's prediction theory. They computed correlations for pressure and temperature data for stations scattered over large areas. Also they investigated correlations existing between various elements in the upper air and the surface. They found that correlations varied appreciably from season to season and also from year to year within a particular season. As a result they could not get a handle on the dynamics of the situation. The variability prevented them from finding a way to get good prediction results from Wiener's theory.

In hindsight, Wadsworth's efforts were preordained to fail. The reason is that weather is produced by a worldwide moving system of oceanic and atmospheric effects. The weather stations available at that time were too few and far apart to effectively sample such a complex natural system. Balloons would be needed to measure weather conditions at altitude. Worse still, the desk calculators could not efficiently keep up with even the sparsity of the data available. The weather stations would have to be many more in number and the observations much more closely spaced in time in order to get good correlations.

In about the middle of 1943, Wadsworth concluded that forecasting might be carried out successfully by attempting to obtain the dynamic law itself from past weather maps. He turned to a method that was a variation of the analog technique originally introduced by Dr. Irving Krick.

Caltech Professor Irving P. Krick

Irving P. Krick (1906-1996) was an American meteorologist and inventor, a controversial pioneer of long-term forecasting and cloud seeding, and a brilliant salesman who in 1938 started the first private weather business in the United States. In 1930, Krick began studying at the California Institute of Technology (Caltech) and completed his doctoral degree in 1934. He remained at Caltech and started the meteorology department with him as head. Krick became famous by asserting that the 1933 crash of dirigible USS Akron was due to an erroneous forecast by the Weather Bureau. Krick was a natural prcmoter and he determined to make money in weather forecasting.

Hollywood was in the process of filming of the movie "Gone with the Wind." A major scene in the movie was the burning of the city of Atlanta, Georgia, during the Civil War. The Hollywood studio could not willy-nilly start a great fire for the movie scene. Heavy winds could spread the fire widely. The studio was desperate to obtain a correct forecast of the weather conditions that would allow such a fire. Krick came to the rescue. He specified the day anc correctly predicted the weather. In this way he seized the Hollywood movie industry for all things regarding the weather.

When Colonel Hap Arnold was stationed at March Field not far from Caltech, Krick became his friend. Krick offered courses for Arnold's nascent Air Force Weather Service. Hap Arnold, promoted to General, became chief of the U.S. Air Forces during World War 2. General Arnold recruited Krick into the Air Force Weather Service. Scientifically inclined meteorologists looked for causes of natural phenomena. In contrast Krick made use of historic patterns and cycles. He reused old weather maps that resembled current situation, arguing that future weather developments will most likely follow the recorded patterns. His simple methods and forceful salesmanship did not endear Krick to scientific elite. However, with General Arnold's backing, Krick was almost unbeatable.

In 1944 Krick was with the U.S. Strategic and Tactical Air Force who were engaged in predicting the weather for the upcoming Allied Normandy Landings. Also involved were the weather services of Britain, the Royal Navy, and the U.S. Navy. Observations from Newfoundland taken on May 29 reported changing conditions that might arrive by the proposed invasion date June 5, 1944. Based on their knowledge of English Channel weather, the British predicted the stormy weather would arrive on June 5. They argued that the invasion should be postponed until June 6. Krick used his analog forecasting method, which was based on historic weather maps for the past 50 years. Krick's forecast was that a wedge of high pressure would deflect the advancing storm front and provide clear, sunny skies over the English Channel on June 5. He argued that D-day should not be postponed.

As the countdown took place, tension between Krick and the British increased. Professor Krick was used to having his own way in Hollywood. Would it be the same with the British? A decision had to be made. The troops were waiting. General Eisenhower's chief meteorologist was British Group Captain James Martin Stagg. Stagg had the courage to make the decision. Because of uncertain weather conditions, Stagg advised Eisenhower to postpone the invasion of Normandy by one day from June 5, 1944, to June 6, 1944. Eisenhower rejected Krick's advice for June 5 and took Stagg's advice for June 6. Captain Stagg proved to be right.

Analog weather predictions

In 1943 Wadsworth, Bryan and Gordon felt that it might be possible to further develop Professor Krick's idea into a useful tool for making long-range forecasts. Thus, by using the analog concept one might obtain a system of cataloging the types of weather phenomenon so that the dynamics of the future could be taken as equivalent to the dynamics demonstrated by a suitable analog. For the analog technique to be useful it must forecast a large-scale movement of air masses, highs, lows, etc., over an extensive area so that it will have dynamic soundness. The best way, of course, to accomplish this would be to include the whole Northern Hemisphere and attempt to find analogs for the complete Hemisphere itself. They divided the Northern Hemisphere into six zones. These zones were chosen in the band of the Westerlies so that they would, as closely as possible, represent the active movement of the weather phenomena. Caltech recorded the pressures for each day from 1899 to 1938. In 1944 and continuing through 1945, Wadsworth, Bryan and Gordon made a large number of studies devoted to the possibility of forecasting by means of analogs. The analogs were picked for the individual zones daily, and usually not more than two were chosen for transmission. The analogs were sent to Washington and from there distributed to both the European and the Asiatic theatres of operation. Correlations were computed between the present map and all of the past maps belonging to the corresponding time of year. A tremendous amount of mechanical sorting and a great deal of tabulator calculation was necessary in order to survey the past records and pick out the previous situations which looked analogous to present weather developments. At no time were very startling results obtained. The reason was as before. The weather stations were too few and far apart and the desk calculators could not effectively do the job.

BELL TELEPHONE LABORATORIES
INCORPORATED
MURRAY HILL LABORATORY
MURRAY HILL, NEW JERSEY
SUmmit 6-6000

May 22, 1953

DR. ENDERS A. ROBINSON
Geophysical Analysis Group
Department of Geology and Geophysics
Massachusetts Institute of Technology
Cambridge 39, Massachusetts

Dear Dr. Robinson:

     Thank you for your letter of May 19. If you come through New Jersey in June there is a fairly good chance that I will be here, but I expect to spend most of July and August at Mattapoisett near the Cape Cod Canal. I shall be staying with my parents and we are the only Tukeys in any southeastern Massachusetts telephone book.

     I look forward to seeing you somewhere.

     With best regards,

Very sincerely yours,

John Tukey

1445-JWT-MAR   J. W. TUKEY

Chapter 12. Discovery of deconvolution

> Henri Poincare: Experiment is the sole source of truth. It alone
> can teach us something new; it alone can give us certainty.

Situation in exploration in 1950

Professor Robert Shrock (*Geology at MIT 1865-1965*, Volume 2, MIT
Press, 1982) describes the state of petroleum exploration in 1948 as
follows:

> Exploration for oil and gas, a vital petroleum industry activity,
> depends heavily on the seismic process for basic data that are
> necessary to determine the attitudes, shapes, physical
> properties, and structural relations of subsurface rock masses.
> These determinations are of critical importance because they
> provide otherwise unobtainable evidence that local subsurface
> conditions may be either favorable or unfavorable for the
> accumulation of gas and oil. Unfortunately the seismic process
> is often rendered useless, or nearly so, by so-called "noise," that
> is everything that appears on a seismogram except the desired
> signal. The ultimate problem for the exploration geophysicist,
> therefore, is to identify and delineate the desired signal that is
> concealed in the background noise.

In the late 1940s, as the country was settling down after the
end of World War 2, but while car pools were still common,
numerous MIT professors living in Lexington rode back and
forth from Cambridge to Lexington in such pools. In one of
these groups were George P. Wadsworth (Mathematics S.B.
1930; S.M. 1931; Ph.D. 1933), a professor of mathematics, and
Patrick M. Hurley (Geology Ph.D. 1940), a professor of geology.
On a number of occasions I also rode in this particular group
and learned firsthand how widely the conversation spread
throughout the general area of instruction and research in
science and technology.

One day in 1948, the exact date is no longer remembered by the participants, Wadsworth and Hurley fell into conversation about the possible use of mathematics in geophysics. Actually, Wadsworth was needling Hurley a little bit; chiding him because neither geologists nor geophysicists were using any refined mathematical techniques at the time–perhaps a little statistics in sampling problems and ore deposit evaluation, some geometry and trigonometry in surveying and crystallography, and a little calculus and some differential equations in representing some simple dynamical situations, but that was about all.

They began immediately to think about organizing a research program and recruiting graduate students to help with the computations. Inasmuch as some poor quality seismic records would be needed for the suggested investigations, an inquiry was sent to the Magnolia Petroleum Company in Dallas, Texas, for such records. Through the cooperation of Dayton H. Clewell (Physics S.B. 1933; Ph.D. 1936), then in charge of the Company's research laboratory, and Jacques Yost, one of their geophysicists, we obtained eight seismograms in the spring of 1950.

Wadsworth had been a student of famed MIT mathematician, Norbert Wiener, and had used some of his defined statistical properties of time-series analysis to develop weather-forecasting techniques for the Fifth and Eighth U. S. Air Force units in the European Theater during World War 2.

By the summer of 1950 Wadsworth had interested one of his students, Enders A. Robinson (Mathematics S.B. 1950), and had induced the Department of Mathematics to employ him as a half-time Assistant for SY 1950-51. He would spend half his time on regular graduate work and the other half, about 16 hours per week, on his assistant's duties. These duties would consist of about 8 hours a week teaching a section in freshman calculus and the other 8 hours on assisting Wadsworth on fundamental geophysical research.

The work of the geophysicist is to look beneath the surface of the earth in the search for deposits of oil and natural gas. The subsurface is made up of layers of rock (much like a layer cake). A geologic interface is the common boundary of two layers such as sandstone and shale. The subsurface geologic structures of interest can be as deep as three or four miles. The oil is trapped within certain layers.

The seismic reflection method for petroleum exploration came into general use in 1930. A seismic survey consisted of collecting data over a selected geographic area, called the prospect. At a fixed point on the surface of the earth, a dynamite charge (a.k.a. the source or shot) is detonated. Seismic waves from the shot propagate downward from the source point deep into the earth. The waves are reflected from the geologic interfaces. The reflected waves propagate upward from these interfaces. A primary reflection is a reflection that travels directly down to the interface, and then directly back up to the surface. A multiple reflection is a reflection that bounces back and forth among various interfaces as it proceeds on its trip. The reflected waves, both primaries and multiples, are detected on the surface by receivers (geophones). The receiver points are located at various horizontal distances from the source point. The signals received at each receiver point for a given source point are recorded (in real time) as traces on photographic paper (called the seismogram or seismic record).

After one shot is completed, the source point and the corresponding receiver points are moved so that another shot can take place. This acquisition method is repeated again and again until the entire prospect is covered. Within the confines of a given exploration budget, a fixed number of source and receiver points must be used. The points are chosen so as to obtain the best possible representation of the prospect.

The geophysicist had to examine the traces by eye in order to mark the primary reflected events. Each primary reflection defines an interface.

The interfaces are plotted to show maps of the cross-sections of the subsurface. The maps are then interpreted by geologists and geophysicists in order to choose the most favorable drilling sites for new oil wells, either wildcats or field-extension wells. Despite these advances, many, in fact, most of the great potential oil producing regions of the world had to be abandoned as active areas of exploration. Seismic records from such areas were classified as NR (No Reflections) because the raw data did not reveal primary reflections. Despite great efforts in instrumentation, encouraging results from the use of various types of analog filters were not forthcoming.

The depths of the geologic interfaces between the major subterranean rock layers could be determined by drilling. However, many of the reflections from these interfaces could not be seen on the records. The oil companies would use various analog filtering methods, but these methods were largely unsuccessful in the detection of these unseen reflections. The problem was what could be done on the NR records to make the unseen seen.

The assignment for the research assistantship was to analyze the eight seismic records of the Magnolia Petroleum Company. Professor Wadsworth and Professor Wiener would supervise the work. Exploration seismic records were not like weather records. In meteorology the weather records extend over many past years. One wants to predict things such as daily temperature and precipitation. In contrast, an exploration seismic record extends over a time span of five or six seconds. You have the entire record before you, so there is nothing to predict. The assumption was that the record was made up of signal and noise. If you could remove the noise, then you would obtain the signal. The signal would reveal the dynamics of the situation. The words signal, noise, and dynamics were used in a general sense, but as to particulars these words were largely unspecified.

Weather is determined by the movements of fluids (water and air). Weather events are here today and gone tomorrow. Seismic waves are determined by the fixed geologic structure. The interfaces between the layers produce the primary reflections. Often the record does not show all of the primary reflections. Just because they are unseen dos not mean the primary reflections are absent. The unseen reflections are indeed present, but they are hidden from view by various types of noise, including multiple reflections. The outstanding problem was to find a way to uncover the hidden primary reflection events on NR records.

MIT Professors Paul A. Samuelson and Robert Solow

In the spring semester of 1951 Prof. Wiener went to Mexico, and he moved me into his office. Alone I missed the companionship of my office mate Chester Gordon. Instead Wiener's books and mementos became my office mate. I was studying under Professors Paul A. Samuelson and Robert Solow who were experts in time series analysis. Later each of them would win the Nobel Prize. Paul A. Samuelson was awarded the Nobel Prize in Economic Sciences in 1970 for the scientific work through which he has developed static and dynamic economic theory and actively contributed to raising the level of analysis in economic science. Robert M. Solow was awarded the Nobel Prize in Economic Sciences in 1987 for his contributions to the theory of long-term macroeconomic growth. Both Samuelson and Solow were outstanding mathematicians, and could just as well have been in the Mathematics Department.

In the academic year 1950-1951, I took Samuelson's advanced graduate course on economic analysis. I also was writing a master's thesis under Prof. Solow. In the spring semester, Samuelsson lectured about the many interesting and fruitful aspect of dynamics which have nothing to do with the business cycle as such, but which are important for the understanding of processes usually classified under the heading of economic theory.

In comparison to the business cycle, the geophysical problem seemed simple. In the children's game of hide-and-seek, the player designated as being "it" does not look while the other players hide. The player who is "it" has to locate all the concealed players. It is not difficult because the hidden players are not allowed to move. The detection of reflections is much like a game of hide-and-seek. On the seismic trace, various reflection patterns lie hidden but they do not move. They are at the same place next week as this week. You are looking for a fixed solid target. On the other hand, weather patterns lie hidden and they keep moving around from day to day. You are looking for a moving fluid target. The business cycle is much more like the weather than the solid earth.

At Harvard University, Paul Samuelson had studied under Joseph Schumpeter and Wassily Leontief. Schumpeter was known for making his students work with a heavy load. After Harvard, Samuelson moved on to MIT. Although Schumpeter encouraged some young mathematical economists, he was not a mathematician, and tried instead to integrate history and sociological understanding into his economic theories.

In class in the spring semester of 1951, Professor Samuelson went over Schumpeter's model of economic development. According to Schumpeter, innovations consist of such things as the introduction of new products or new methods of production. The entrepreneur is not an ordinary manager, but a person with energy and ingenuity who introduces something entirely new. He is driven by a creative impulse. When the entrepreneur introduces a new innovation for the purpose of earning profit, the existing economic flow is jolted. Once the new innovation becomes successful and profitable, other entrepreneurs reinforce it. For example, the emergence of computers stimulated a wave of new investments in other things. Let us make an analogy. The jolt of a new innovation is like a pebble dropped into a puddle, and the wave of new investments is the ensuing wavelet. On this basis I worked out a mathematical model for the Schumpeter theory of economic innovations. I converted Schumpeter's verbal analysis into equations, using Wiener's prediction theory in this endeavor.

The geophysical problem was to find the hidden reflections in the seismic trace. The economic problem was to find the innovations in the underlying economic time series. The crucial link was the recognition that the onset of an innovation, such as a new technological advance, cannot be predicted. Thus the onset of each innovation produces a definite and measurable prediction error.

On this basis, the timing of the economic innovations can be found from the economic time series as follows. First compute the prediction operator for the given time series. Next apply the prediction operator to obtain the predicted values. However, the predicted values are not the object of the analysis. Instead, the prediction errors are desired. The prediction errors can be obtained by subtracting the actual values of the time series from the predicted values. These prediction errors represent the desired economic innovations.

Could the same method work in exploration geophysics? My plan was to take the digitized seismic traces and treat them as economic time series. I would carry out the prediction error filtering.

Detection of reflections

In 1950 the seismic observations were continuous in time and the seismic stations were closely spaced in distance. All of this boded well for the seismic. The seismic method worked in various cases. The problem was no one of discovering a new way of doing things. It was a way of improving an existing system. More particularly, the seismic method worked in those cases when the reflections could be seen by the naked eye. In other cases, we could assume that the reflections were indeed present, but they were hidden from the naked eye. The problem was clear. A computational method had to be found that would strip the interfering things away from the reflections in order to make them visible. Dean Clark (*The Leading Edge*, February 1985) writes:

> The seismic records had to be initially digitized by hand measurements; the numerical filter was to be applied via a hand calculator; and the filtered numbers then re-plotted by hand. "It was a lonely feeling, especially at MIT at night, digitizing the records with a T-square, ruler, and pencil," Robinson says. "Except for Wadsworth, nobody at MIT or in the entire oil

industry thought the analysis of digital seismic data would ever be feasible." Robinson wrote down read ngs of the seismic trace at intervals of 2.5 milliseconds, in total 600 to 800 readings per trace. Virginia Woodward, one of a platcon of women working under Wadsworth who were expert in operating a desk calculator, then performed the fairly simple but extensive numerical filtering operations. Several weeks later, a new batch of numbers [the ones from the calculations] was returned to Robinson who re-plotted the filtered traces, the first products of the process now known as *deconvolution*. (The hand-deconvolution of 32 traces took the entire summer of 1951. Currently [i.e., in 1985], some oil company computer systems regularly deconvolve more than a million traces a day.) When I started plotting those first traces, I wasn't expecting anything," Robinson recalls. "We had tried other things and hadn't found any pattern. There wasn't any reason to think this process would be any better. I couldn't imagine we would ever be able to pick out anything from a seismogram "

The first hand-plotted trace, though, was a revelation. The good reflections from the original record showed up, and some that weren't so strong also came through. Robinson was initially amazed but quickly deemed the result a fluke. He was certain he would never see another as good. But the second trace, then the third, confirmed that the first had been no accident. It was impressive empirical evidence of a revolutionary discovery – numerical filtering could separate data and noise just like electronic filtering.

Robinson's first duty, of course, was to inform Wadsworth. However, in 1951 that was a difficult chore, particularly for a first-year graduate student. At that time Wadsworth was almost completely inaccessible. After World War 2, he had emerged as one of the busiest and highest-priced consultants in the United States. His schedule was so crowded with high-ranking visitors from industry that appointments, usual y for just 5-10 minutes, had to be made weeks in advance. Although Robinson's request obviously had some urgency, he could not get in to see Wadsworth for three weeks. [It was the end of August 1951].

The long awaited meeting was not, to Robinson's surprise, a celebration. "It was very disconcerting," he says. "It was as if you had found a gold mine in your backyard, filled your pockets with gold, and then had your boss say, "That's nice - but you should really be looking for lead." To me, what we had found looked like magic. But Wadsworth wasn't really that interested in seismology. He wanted me to come in with equations." Geologist Hurley, however, was ecstatic and he immediately initiated action to obtain financial support from the petroleum industry [specifically Magnolia Petroleum Company] for more extensive research.

Magnolia Petroleum Company was initially cautious, for at least two good reasons. First, Robinson's technique, which involved breaking the data down into discrete samples and using only a short section of the trace to analyze the rest, was opposed to the conventional industry wisdom. Second, even if the method did obtain good results, at the time it could not be performed economically. Before committing itself to a full-scale research effort, industry had to be assured that instrumentation could be developed which would accelerate the process.
By remarkable coincidence, Robinson got an immediate opportunity to respond to industry's concern over instrumentation. The chance came because, just as Hurley was hustling oil company support, MIT announced that its new Whirlwind digital computer would, for the first time, become available for use by the academic community.

A seismic reflection is a physical phenomenon. It is governed by Newton's law of motion. A prediction error is a mathematical entity. It is governed by mathematical equations. The finding was that a physical phenomenon (seismic reflection) was connected to a mathematical entity (a prediction error). This finding was a major discovery. All else would rest upon its validity. The finding could be a fluke. It could be an accidental alignment that held true only in scme particular case. Two courses were open, one was empirical confirmation and the other was mathematical confirmation. Empirical confirmation would involve the validation of this result by computing a lot of other cases. Such confirmation would be more convincing to the practical person because, like-it-or-not, "seeing is believing." Concrete physical evidence is convincing. Mathematical confirmation would take longer. The first question to be answered was: What was there in Wiener's mathematics that made such a physical result possible. What were the underlying physical assumptions?

Funds had been available for only one year of seismic research. As of September 1951 the time was up. It was sink or swim. The Mathematics Department transferred me to weather prediction. The first thing was to compare the problem of weather prediction to the problem of detecting seismic reflections. In 1951 the weather observations were meager and the weather stations were widely space in distance. Everything in weather is continually changing. The atmosphere and oceans are huge masses of continually moving fluids. In effect, weather prediction is concerned with a fast moving target of worldwide scope.

The problem of seismic exploration was easy in comparison. The earth's upper crust is indeed moving over the millions of years of geologic time. However it may be considered as fixed, solid and unchanging in the time span of a human being. In this respect, seismic exploration involves a stationary target. A seismic prospect is concerned with only a small section of rock layers which are immoveable. A seismic record taken on a given day is identical (in all practical respects) to a seismic record taken in the same place weeks later. Seismic exploration in 1950 was concerned in looking at an area of only few square miles in extent and a depth of one to three miles. Everything in that volume was solid rock. Any water, oil, or gas was tightly held in the pores of the rock.

Chapter 13. Geophysical Analysis Group

James Clerk Maxwell: The dimmed outlines of phenomenal things all merge into one another unless we put on the focusing-glass of theory, and screw it up sometimes to one pitch of definition and sometimes to another, so as to see down into different depths through the great millstone of the world.

Use of Whirlwind in 1952

In 1952, digital computers were in the news. National press coverage made computers a hot topic. The first digital computer was the ENIAC. It was built by J. Presper Eckert and John Mauchly for the Army at the University of Pennsylvania. Eckert and Mauchly left the University in March 1946 and formed a company to build computers. The result was the UNIVAC. It was the first American computer designed for business and administrative use. Mercury delay lines were used to store the computer's program. The binary digits circulated within the delay lines in the form of acoustical pulses that could be written into the line and read from it. The first UNIVAC was delivered in 1951 for the U.S. Government's Census Bureau. The first commercial customer to purchase a UNIVAC was the Prudential Insurance Company. By the summer of 1952 there were many news items about the ENIAC and the UNIVAC. The general public was becoming aware of computers. A computer was supposed to be able to do anything. A digital computer was heralded as a "Giant Brain." There even were undercurrents that computers would replace the need for humans in many endeavors. Even then it was noticeable that some of the students working on Whirlwind had a compulsive relationship with the computer. They loved the computer. However, the general public had no working knowledge of computers. There was a noticeable anxiety about computers. When asked, most people replied that they would never use a computer.

Professor Robert Shrock (*Geology at MIT 1865-1965*, Volume 2, MIT Press, 1982) writes:

It was largely through Wadsworth's urging that Robinson transferred from the Mathematics Department to the Department of Geology and Geophysics in February 1952, accepted an appointment as Research Associate in the Department of Geology and Geophysics for the spring term of academic year 1951-52, started on a doctoral thesis, and simultaneously began to organize the research program that became the MIT GAG Project in early 1953. Once Robinson got fully involved in the research, he quickly found that he needed help, which meant graduate students, computers and secretaries, and of course these would require money."

Thus it happened that Hurley, who was serving with Wadsworth and Bryan as an MIT Advisory Committee for the research, came to me with a request for $13,000 to cover the costs of the first stage of the expanded program of research, to start in the fall of 1951 under Robinson's direction. The request was immediately taken to George R. Harrison, then Dean of the School of Science, with an explanation of what we wanted to do. He quickly grasped what Wadsworth and Hurley had in mind and in his typical manner, somehow and from somewhere, he got the money for us. [Added note: The money came from the office of MIT President James R. Killian.] This was neither the first time nor the last that Dean Harrison, a good friend and strong supporter of our Department in those days, found the funds to help Geology; unfortunately, Deans, and sometimes even Department Heads, do not get the credit they deserve for finding that all-important 'seed corn' money when most needed. Herewith I am happy to record our gratitude to Dean Harrison. Thus the $13,000 became available in February 1952, and we could move to the next stage.

In their earlier experiments Wadsworth and his colleagues had found that the necessary computation was expensive, so it was clear that if a major computation program was mounted funds of the order of $50,000 would be needed. Helpful as the

Institute's $13,000 would be, it could really only serve to get the program started. Where, then, could Wadsworth, Hurley, and the rest of the GAG seek such funds? It seemed obvious that the most logical group to approach would be the petroleum companies and their geophysical subsidiaries, since they would be the ones presumably to profit from any useful results that might come from the proposed research. At this juncture the research group found a staunch supporter in another MIT alumnus, then Geophysical Research Director of the Stanolind Oil and Gas Company (now AMOCO Production Company) of Tulsa, Oklahoma [Added note: later acquired by BP]. This was Daniel Silverman (Electrical Engineering, S.M. 1929; Sc.D. 1930). Being an electrical engineer and well trained in communications theory, Silverman thoroughly understood what Wadsworth and his group had in mind and enthusiastically joined them in inviting the major petroleum companies to send a representative to a proposed information meeting.

With fund-raising in mind, arrangements were made with the MIT Industrial Liaison Office to present the seismic research results to a group of invited petroleum and geophysical companies on 6 August 1952. At this meeting, which was held in Cambridge, Wadsworth and Robinson presented a paper on "Results of an auto-correlation and cross-correlation analysis of seismic records."

At the August 6, 1952 meeting, it was not the mathematical equations that impressed the attendees. Instead it was the computer-processed traces on the Whirlwind. The geophysicists showed a great enthusiasm. All that they knew about digital computers was what they had read in the newspapers. None had ever used a digital computer. Some had spent years in swamp and field searching for oil. Others were working on the development of geophones and recording instruments for seismic exploration. They were all dedicated to the development of the seismic exploration.

In the story *The Wonderful Wizard of Oz*, Dorothy sets off to see the Wizard. There were several roads nearby, but it did not take Dorothy long to find the one paved with yellow bricks. Within a short time she was walking briskly toward the Emerald City; her Silver Shoes tinkling merrily on the hard, yellow road-bed. At the end of the road Dorothy did find the Emerald City. Our message was: The digital computer is the yellow brick road. Follow the yellow-brick road and it will lead you to the Emerald City of petroleum.

There was one great omission at the 1952 meeting. The attendees were not taken to the MIT Barta Building. There they would have seen Whirlwind, as Professor Wiener would say, "in the metal." The machine was an intricate work in progress, a huge concoction of electronics, vacuum tubes, and flashing lights. It was hypnotic. Would they have fallen under its spell? Deconvolution was successful on the Magnolia seismograms. However, one flower does not bring the spring. Our work was cut out before us. Many more seismic traces would have to be processed before it could be said that the deconvolution gave the the reflecting geologic interfaces. A firm link had to be established between the mathematics and the geology.

Formation of Geophysical Analysis Group in 1953

Professor Robert Shrock (*Geology at MIT 1865-1965*, Volume 2, MIT Press, 1982) writes:

> Soon after the meeting on August 6, 1952, the Department of Geology and Geophysics was authorized to solicit financial support from a number of companies and to organize and carry on a major program of research using time-series analysis on seismic records. With the assistance of the Industrial Liaison Office, particularly of Robert V. Bartz, and with the energetic help of Daniel Silverman, a total of $52,300 was obtained from 13 companies as shown on the accompanying financial record.
>
> With an initial total commitment of $52,800 by the 13 companies Wadsworth and his colleagues could now proceed with already developed plans, and the MIT Geophysical Analysis

Group, which had actually been developing during the summer and fall of 1952, officially came into existence in February 1953. Enders Robinson, as previously mentioned, had been directing the research project since 1950 and was designated Director of the GAG when it was officially established. He would serve ably in this capacity until September 1954, when he would leave MIT with his Ph.D. degree in hand and be succeeded by Stephen M. Simpson, Jr. who likewise would ably serve until termination of the project in June 1957. Wadsworth, Hurley, and Bryan became an MIT Advisory Committee, and Silverman generously agreed to serve as Chairman of the Industry Advisory Committee, a service that was of inestimable value to the GAG Project throughout its entire existence. Furthermore, he has repeatedly publicized the great impact that the MIT GAG Reports have made on exploration geophysics and the petroleum industry.

During 1952, as seismic research went forward largely under Robinson's direction, and as plans were being developed in anticipation of the GAG Project, a small group of graduate students was recruited to serve as research assistants. S. M. Simpson, Jr. joined the group early in 1952, to be followed in September by H. W. Briscoe, M. K. Smith, and W. P. Walsh. Simpson quickly assumed an important role as an assistant to Wadsworth and Robinson.

With the formal establishment of the Geophysical Analysis Group in February 1953, with Robinson as Director, things really began to happen. Space had been secured on the second floor of Building 20E, student assistants had been appointed, additional seismic records had been obtained from several companies other than Magnolia (The Texas Co., Atlantic Refining, Cities Service, and Amerada), and the overall program rapidly got underway. Thereafter there were many meetings of the MIT staff members and regular semi-annual meetings at which latest results were reported to representatives of the supporting companies, including members of Silverman's Industry Advisory Committee These were exciting days for the participants in the GAG.

Enlistment of Raytheon

During the years 1952 and 1953, hardware improvements were being made on Whirlwind. The machine had a lot of down days. At that time, the maintenance time on Whirlwind's memory alone was four hours per day, and the mean time between memory fa lures was two hours. Whirlwind was a military computer. Time was allocated to the MIT faculty and students. There was no way that any outside organization would be allowed to use Whirlwind. The GAG needed to find a commercial computer service so that the oil companies could do digital signal processing right away as a prelude to setting up their own in-house operations.

The ENIAC was the progenitor of all electronic digital computers. The ENIAC served as the mainstay for scientific computation during the period 1946 through 1952. It surpassed all other existing computers put together whenever it came to problems involving a large number of arithmetic operations. With the end of the war in 1945, the computation of ballistic tables for the U.S. Army was no longer needed. The first computation on the ENIAC was done by John von Neumann. The results verified that the Teller-Ulam design of the super bomb was feasible. During its lifetime, the ENIAC was used for all kinds of scientific computation including weather prediction, atomic-energy calculations, cosmic-ray studies, thermal ignition, random-number studies, and wind-tunnel design. No electronic digital computers were applied to commercial problems until about 1951. EDVAC and ORDVAC, both faster than ENIAC, began to share the work load with the ENIAC in 1953 at Aberdeen.

The mathematician behind the ENIAC was Professor John von Neumann of the Institute of Advanced Study. Great in every respect, he made vital advances in mathematics, quantum theory, nuclear physics, computer science, and econometrics. Early in his life he did important work in set theory, algebra, and quantum mechanics. At the age of 20, he published the definition of ordinal numbers that is still used today. He also developed the "theory of automata," which compared the computing abilities of a machine with that of a human brain. He devised the computer infrastructure known as the "von Neumann Architecture."

On all things regarding the ENIAC, John von Neumann worked closely with Dr. Richard Clippinger. Clippinger was in charge of the ENIAC at Aberdeen from its inception in 1946. Clippinger was also responsible the development of the EDVAC and ORDVAC computers at Aberdeen. In 1952 the Raytheon Manufacturing Company in Waltham, Massachusetts set up a computing group. Raytheon employed Clippinger as its head. Clippinger had a PhD in mathematics from Harvard. With Clippinger came his colleagues Bernard Dimsdale and Joseph H. Levin, each with a PhD in mathematics. Before entering Raytheon, Dimsdale was responsible for numerical analysis at Aberdeen; and Levin was responsible for the SEAC computer at the National Bureau of Standards.

Franz Leopold Alt (1910 –2011,) was an Austrian-born American mathematician who made major contributions to computer science in its early days. He was best known as one of the founders of the Association for Computing Machinery, and served as its president from 1950 to 1952. In World War 2, he served in the elite 10th Mountain Division, trained for skiing, rock climbing and mountain fighting. In 1945 the Army put Alt in charge of planning for electronic computation at Aberdeen Proving Ground. As a civilian he returned to Aberdeen as Deputy Chief of the Computing Laboratory. In 1948 he transferred to become Deputy Chief of the Computation Laboratory at the National Bureau of Standards in Washington, DC, where he directed the early use of computers in the federal government. In his Oral History Interview by the American Institute of Physics on February 24, 1969, Franz Alt said:

> Now, when I finally got to Aberdeen the second time in 1947 —
> when I physically moved to Aberdeen, meanwhile ENIAC had
> been installed there, and I began to be much more interested in

that. And, just at that time the principal interest was in changing over ENIAC from plug-board programming to punched-card programming, and that made a great deal of difference in the use of the machine. ENIAC also would have been obsolete from the time it was built except for two things: One was that the other electronic computers, real stored program computers, were still slower in coming — were not finished — and the other was that von Neumann invented the method of programming ENIAC by means of punched cards; and that gave it a new lease on life. The old way, where each new program as set up by plugging wires into panels (and there were I don't know how many hundreds of wires to be plugged for each program, the panels filled the whole room) it took days to change over from one problem to the next. It was quite like a DIFFERENTIAL ANALYZER, which also took several days to set up. ... But we began to see gradually that this did not take advantage of the speed of the machine. ... **John von Neumann** invented a new system quite comparable to today's automatic programming. Instructions were punched into cards, the cards were read. The way this machine was designed, cards would be read only for input data in the course of a problem. But, here now, the machine read the cards, stored the numbers as numbers, and then interpreted these numbers as instructed. And the machine was permanently wired to interpret each number as an instruction. And that so-called 'New Programming Method' was being designed — it was von Neumann's idea, but it was **Richard Clippinger** who mainly implemented it, together with **Bernard Dimsdale**, those two — and it took probably a year or so before it was all tested out. Toward the end of that year, which might be in the spring of 1948, we began to set up real problems by this method. ... Yes, the DIFFERENTIAL ANALYZER [at Aberdeen] was a section by itself. **Dr. Joseph Levin** was in charge of that and I knew him many years after that. He came to the Bureau of Standards with me and then he went to Raytheon at Waltham.

Here was a chance for the GAG to have the three mathematicians Clippinger, Dimsdale and Levin. They successfully implemented the computational ideas of John von Neumann.   What better recommendation could there be? And best of all, Raytheon was as close to MIT as any corporation could be. I contacted Raytheon and on October 28, 1952 Raytheon submitted an estimate. Raytheon was anxious to do computer work for the GAG. In this way Raytheon would be in a position to offer services, backed by pertinent experience, directly to the oil industry. In addition Raytheon would be sufficiently familiar with the problem to design especially suitable computing equipment including input and output devices.

Computations in the spring of 1953

By February 1953 the money necessary for GAG consortium was fully obtained. Invitations to join the GAG had been sent to the major American oil and geophysical companies. All accepted except two. Henry Salvatori, who was head of Western Geophysical Company, declined. Shell also declined, but not on the merits of the project. It was Shell's policy not to engage in any joint endeavor with any other oil company. In fact, this policy more or less held true for all oil companies. The GAG is the first instance when nearly all the members of a major industry came together to work on a research project. The MIT Industrial Liaison Department deserves the credit.

The term "dry hole" is used in oil exploration to describe a well where no significant reserves of oil are found. This term is now often used to describe any fruitless commercial initiative. In the early 1950s the seismic method was rudimentary. The seismic traces were recorded on paper. The traces were examined by eye and the supposed reflections were marked by pencil. From these readings a contour map was hand drawn using a large amount of imagination. The maps were handed to the geologist, who would mark the drilling sites. The geologist was never told what geologic features might be figments of imagination.

That did not matter, because the company would keep drilling more holes until oil was found or until they gave up.

Spending money for dry holes was not the major problem. The big impediment was that many of the possible oil bearing regions of the world yielded seismic records that had no visible reflections. With no clear geologic indication, an oil company had no idea on what exact spot they should begin drilling. For this reason great areas such as the Gulf of Mexico and the North Sea were left untouched.

The immediate goal of the GAG was to compute further examples of successful deconvolution. It was necessary to give more demonstrations that deconvolution yield the locations of the subsurface geologic interfaces. The plan was to show that deconvolution worked on any seismic record. Whirlwind was a military computer that was unavailable for use by the oil companies. Raytheon was a computing service that any oil company could use on an independent confidential basis.

The winter meeting of the GAG was held in Dallas on January 30-31, 1953. Specifically, the GAG proposed to do about 60 cases each month to verify the validity of deconvolution in various geologic settings. Raytheon would do the computations. It would be paid for by money already contributed to the GAG. The 14 oil and geophysical companies present agreed to go ahead with the proposal. Plans were discussed. The oil companies would provide various assortments of seismograms to be used. On February 2, 1953 the GAG held a meeting with Raytheon representatives. A few days later Raytheon sent prices for the work involved in the programming, the digitization of the records, the computation, the supplying of operator coefficients, and the final charting of the error curves. Raytheon would do a total of 320 cases represented about five months computational work at the rate of about 60 cases per month. Thus 320 cases would cover the months of February through June 1953.

Raytheon decided to use a machine at the University of Toronto (UT) for the deconvolution computations. The Toronto machine was called

FERUT (for Ferranti at UT). Raytheon selected this machine after a careful survey of the five or six large scale electronic digital computers available for commercial computations. The Ferranti computer incorporated the genius of Alan Turing.

Dr. Alan Turing and the Ferranti computer

Alan Turing (1912-1954) was a British mathematician who covered a whole spectrum of subjects, from philosophy and psychology to physics, chemistry, and biology. He combined high-level thinking with hands-on experience with machinery and experiments. He is often considered to be a father of modern computer science. A brilliant original thinker, Turing published, "On Computable Numbers" in 1936, describing what came to be called the "Turing Machine." In essence, the Turing machine is an abstraction that could, in principle, solve any mathematical problem described in symbolic form. In 1937 Turing and John von Neumann had their first discussions on what would later be called "artificial intelligence" (AI). In 1950 Turing developed the Turing Test. It deals with a machine's ability to exhibit intelligent behavior. Specifically it gives a way to find out whether a machine is conscious and can think. You would listen to conversations between a human and a machine. If you could not tell the machine from the human, the machine would pass the test. In 1952 Kurt Vonnegut published the novel "Player Piano" set in the future. It is the chilling tale of an engineer who rebels against a world dominated by a supercomputer and run completely by computer controlled machines.

During the World War 2, Turing served in the British code-breaking department. The Enigma was a machine used by the Germans to encipher their military and naval signals. Turing developed a machine called the Bombe that was capable of breaking the Enigma encryption.

At the University of Manchester, British engineers Frederic Williams and Tom Kilburn developed the Williams tube. It was a cathode ray tube adapted to electronically store to store binary data. It was the first random-access memory (RAM) digital storage device. By June 1948 they had incorporated it in a small electronic computer called the Baby computer. The University created a Computing Laboratory and appointed Alan Turing as Deputy Director to draw upon his knowledge to build a full-scale version of the Baby computer. (Figure 7.) In turn, the British government commissioned Ferranti Ltd. to produce commercially version, which was called the Mark 1. , The first Ferranti Mark 1 debuted in 1951. It was the first commercially available general-purpose electronic digital computer. In 1952, the University of Toronto purchased the second Ferranti Mark 1 (the FERUT). It was used to design the St. Lawrence Seaway.

Figure 7. Drum from Manchester-1 computer designed by Turin

The FERUT operated in the binary number system. Its overall dimensions were two bays, each of which was 16 feet long, 8 feet high, and 4 feet wide, and a control desk. The power consumed was 27 kilowatts. The computer had about 4,000 vacuum tubes and 15,000 resistors. Its multiplication time was 2.2 milliseconds (i.e., about 450 multiplications per second). Its addition time was 1.2 milliseconds (i.e., about 833 additions per second). The input was punched teleprinter tape which was read photo-electrically and fed in at rate up to 200 characters per second. The output was either punched teleprinter tape at 10 characters per second or else direct printing by teleprinter at 6 characters per second. It had internal high speed electrostatic storage (RAM) of 256 words (a word being 40 binary digits, which is about 12.1 decimal digits). Its magnetic drum had storage for about 16,000 words.

Turing was mainly responsible for the decision to use the base-32 numerical system for the Ferranti. To do any serious programming, you would have to learn the table of correspondence between the 32 digits of the number base, their numerical equivalent, and their equivalent in binary notation. Both the binary form and the base-32 form of numbers were written with the least significant digit to the left. This convention was for the convenience of the engineers. After the simplicity of Whirlwind, the coding for the Ferranti seemed awful. However, Clippinger was so ingrained with programming skills that nothing was awful to him.

In doing the GAG work in the spring of 1953, Raytheon was plagued with frequent breakdowns of the FERUT. Out of the total time used by Raytheon on the FERUT, 70 percent was no good, for which they were not charged. When the machine went down for a week, the work of Dimsdale, Clippinger and Levin plus the work of two or three woman assistants was wasted for that period. It must be remembered that both the Whirlwind and Ferranti computers were newly designed and were still being tested. This was the first digital signal processing ever to be performed. The whole experience of digital signal processing was new even for Clippinger. The early computers had no floating point arithmetic, so there was always the problem of overflow unless the data were carefully scaled beforehand.

Some of the advantageous aspects of the Ferranti Computer for seismogram analysis work were: a fairly satisfactory input speed, very satisfactory computational speed, and large drum storage. A disadvantage of the machine for this type of computation was its small electrostatic storage (RAM). For an extensive computation, such as the deconvolution, many instructions were required (some 1,000 instructions), as well as large amounts of data. When read in initially all this information was stored on the magnetic drum, and must later be transferred to the electrostatic storage for use in many successive relatively small blocks. The most important consequence of this fact was the difficulty of preparing the code and consequent possibilities of coding errors which must be located and corrected. Another limitation of the machine was its slow output speed. While it was possible to reduce loss of time due to this cause by taking advantage of the fact that the machine could compute during printing, in any specific case other coding considerations sometimes limited the extent to which this could be done.

The Ferranti computer itself was a severely limiting factor. Its layout was not suited for digital signal processing, and yet Raytheon had completed the task. Although both Whirlwind and the Ferranti were general purpose computers, they were quite different with respect to what they could do best. The Ferranti had a long word length, slow speed, and slow output. Whirlwind had a short word length, fast speed, and fast output. Both computers had a small RAM (the Ferranti with 256 long words, and Whirlwind with 1024 short words).

In summary, Raytheon developed a fully operational signal-processing system in five months. According to the IEEE History Center, Raytheon offered to the industry at large what must have been the first commercial digital signal processing service. Never before had computers been used to process large amounts of digital data. Before this time, computers had been used to solve deterministic mathematical equations which involved little actual data.

Figure 8. Alan Turning, John von Neumann, Norbert Wiener

Most significantly the GAG brought together the three giants of early computing (Figure 8):

Alan Turning (in the form of his computer, the Ferranti Mark 1)
John von Neumann (in the form of his computer programmers, Clippinger, Dimsdale and Levin),
Norbert Wiener (as mathematics faculty advisor to the Geophysical Analysis Group)

Deconvolution results in 1953

In 1953 Raytheon reported that twenty-four man months of effort were required to program seismic problem and to check the codes on the Ferranti computer. The Raytheon codes did the same mathematics as the codes that Howard Briscoe and I had written in 1952 for Whirlwind. However the Raytheon codes were much more refined and versatile. The Whirlwind codes were limited by the primitive state of Whirlwind in 1952. At that time Whirlwind had only 1 kilobyte of unreliable electrostatic RAM. It had no magnetic drum. It just had primitive machine language with no floating point arithmetic and no double precision. The Ferranti machine of 1953 was better equipped to handle the processing of seismic records. Whirlwind had the speed, but its short word length was a major disadvantage in solving the mathematical equations for deconvolution.

On July 21, 1953 the GAG submitted deconvolution results for 300 cases in a report to the oil companies. The results looked good. The deconvolved curves brought out the existence of a consistent set of reflections. On the basis of this empirical evidence, digital seismic processing worked. On August 12-13, 1953 a meeting was held at M IT with the oil companies. The geophysicists present were not a homogeneous group. There were field geophysics and there were research geophysicists. Their education was in geophysics, or in geology, or in physics, or in electrical engineering, or in business. Each person had to reconcile the new digital way of thinking with his own experience.

The weak link was that the seismic industry was one of paper and ink. The traces on the paper seismic records had to be read point by point and written down onto paper. The deconvolved traces emerged from the computer as lists of numbers that had to be plotted point by point on paper. The quality of the plotting done by students (with pencils not ink) generally did not meet industry standards. The reaction of the GAG and Raytheon was to let the computer eliminate the paper work.

From a scientific point of view, the computations of Raytheon in 1953 provided essential empirical evidence; namely, the 300 deconvolved seismic traces revealed the physical reflections. This result happened whether the reflections in the seismogram were hidden or not. The reflections are caused by the geologic interfaces between rock layers. The mathematics of deconvolution gave these physical reflections. There was a definite link between mathematics and physics.

Situation in 1953

Dean Clark (*The Leading Edge*, volume 4, 1985) writes

> Obviously, money was industry's major contribution to GAG. It instantly shifted the project's momentum from low to high gear and made the work attractive to some of MIT's brightest graduate students, most of whom subsequently went into geophysics.

> This gave the profession a running start on the coming digital revolution that would hit all scientific fields in the 1960s. However, industry made another vitally important contribution to GAG; input from industry representatives - which included many, such as advisory group chairman Daniel Silverman of Amoco, who were veterans of decades of seismic experimentation - often kept the commercially naive GAG researchers from detouring along unpromising avenues.

The graduate students wanted to write their theses on an assortment of different things. Life was good in the GAG. Spirits were high. The Whirlwind computer was the common thread that bound the GAG together. The Whirlwind digital computer corresponds to "whitewashing the fence" in Mark Twain's story:

Tom Sawyer sets to work whitewashing the fence, but soon concocts a scheme. Tom's buddy Ben Rogers, a notorious teaser, comes by. Ben's enjoying his Saturday, pretending to be a big old steamboat, but when he finally stops and tries to call attention to Tom's troubles, to his work, Tom pretends that he'd rather be painting fences than anything else, even playing or swimming. This makes Ben very interested indeed and in no time – by pretending that fence-painting really is an art – he's got Ben painting the fence and Ben's prized apple in his hands. By the time Ben's tired out, Tom's already got a new mark lined up, and the process continues – Tom taking payment for the "privilege" of whitewashing – until, before too long, the fence is completely painted. Tom, well rested and considerably wealthier – wealthier, that is, if broken glass, old keys and one-eyed kittens constitute wealth. He heads back to report his good work to Aunt Polly.

For thesis work, each GAG member would pursue his or her own line of geophysical research. The only requirement was that they would make use of the computer. In February 1952 Stephen Simpson became the first to join the GAG. He was a graduate student in geophysics who had received his undergraduate degree in physics at Yale. He was a little reluctant at first, but all of a sudden the computer bug grasped him and he became one of the foremost programmers of all MIT, and the strongest advocate of computers. Simpson was thoroughly convinced of the value of the computer in geophysics. He relished in the computer and he freely gave his time in helping GAG members and others in the use the computer. He was a tremendous asset and added much to the enterprise.

MIT Professor Stephen Simpson

Professor Robert Shrock (*Geology at MIT 1865-1965*, Volume 2, MIT Press, 1982) writes,

These were exciting days for the participants in the GAG, the days of 1953 to 1957, and one can get some of that sense of excitement from the unpublished commentary, previously mentioned, that was presented to the Western Geophysical Company Seminar by Stephen Simpson on 7 September 1967, when he recalled his own participation in the program. Following is an excerpt from that commentary that reflects some of the excitement: "The GAG was a terrifically exciting and enriching experience for the graduate students participating."

Stephen M. Simpson, Jr. (PhD Geology and Geophysics, 1953), contributed importantly to our developing program in geophysics during the years 1952 to 1964. ... Only a very few professors [in the 100-year history of the Department of Geology and Geophysics] carried the main load of supervising the thesis work of the SM degree candidates. Four professors— Whitehead, Lindgren, Buerger, and Simpson— supervised 34% of the theses (74 out of 216); the remaining 142 were distributed among 59 individuals, of whom only 10 supervised more than 5 theses, and 21 only a single one.

The following excerpts are from the book *Time-series computations in Fortran and FAP* by Stephen Milton Simpson (Addison-Wesley, 1966).

In the fall of 1952 I joined, as a graduate student, a MIT project called the Geophysical Analysis Group, and so began a twelve-year effort in the application of digital computers to time-series problems. This project, the GAG, was organized by Professors G.P. Wadsworth and P.M. Hurley of MIT and by Dr. Daniel Silverman of the Stanolind Oil and Gas Company. It assumed the task of attempting the realization of Norbert Wiener's time-series concepts on the Whirlwind Computer in the echo-sounding problems of seismic exploration for oil.

At the same time I developed a close friendship with my fellow student Enders A. Robinson, on whom the directorship of GAG soon devolved. Robinson's efforts centered in the elucidation of theory and its translation to discrete notation, and my own work tended toward machine realization of theory, but we each made sufficient excursions into the other's domain to form a profitable research partnership. This pattern has persisted over the years.

The Geophysical Analysis Group is relevant for the reason that many of the programming concepts presented herein were seeded in the 16-bit registers of Whirlwind for the seismic exploration problem. Digital prediction, both single and multiple, special digital filtering, spectral and correlation analysis, traveling spectral analysis, automatic processing systems for multi-trace seismograms, and many other operational concepts were developed and experimented with on Whirlwind to an unprecedented degree.

Besides Robinson and me, those involved with computation included Mark Smith, Howard Briscoe, William Walsh, Robert Bowman, Freeman Gilbert, Sven Treitel, Donald Grine, Kazi Haq, Donald Fink, Robert Wylie, Manuel Lopez-Linares, Richard Tooley, and Robert Sax. The ideas carried into industry and pursued there by students associated with GAG have now ripened to the point of causing what amounts to a technological revolution in seismic interpretation.

In 1954 Robinson left, eventually to become Associate Professor of Mathematics at the University of Wisconsin [in 1961], and I assumed directorship of GAG until its termination in 1957, but frequent visits with each other kept alive our mutual interests.

With GAG's termination and the subsequent retirement of the Whirlwind Computer, I was forced to the realization that my programming output might just as well have been expressed in vanishing ink — an experience which rankled long and which underlies our determination to develop stable programming and communicating techniques.

I took a year's leave of absence from my Assistant Professorship in the Department of Geology and Geophysics at MIT and spent it in military applications of special-design general purpose computers with RCA. This work tended to keep me from recognizing the latent power of the then infant language of FORTRAN.

On returning to MIT, I kept my hand in programming on the IBM 704, but it was not until 1960, when I was asked by the Advanced Research Projects Agency (ARPA, now DARPA) to set up a project like GAG but focused on the underground detection problem of VELA UNIFORM, that I became seriously involved with the new computers. I was fortunate in being able to attract Robinson as a consultant to the project, as well as many

In this congenial and loosely structured group, considerations of programming technique and style were developed to refined levels. Although the authorship of the programs is given individually, I would like to emphasize the importance of the contributions of James Galbraith, Jon Claerbout, and most particularly Ralph Wiggins. Other students directly associated were William Ross, Cheh Pan, Carl Wunsch, and Roy Greenfield.

As for what theory we include in Volume II, much of it is pure review, but some of it has previously appeared only in project report form. I consider Robinson's solution, in the fall of 1962, of the multi-input iteration problem to be a significant achievement. Wiggins pursued and expanded the analysis from this base through the program-development stage, and in so doing was the first to demonstrate the computational feasibility of multi-input least squares.

ADDED NOTE: A small portion of the code from a Whirlwind program written in 1953 is: ta 223, ta 746, ca 746, su 1001, ts 746, sp 625, ca 1010, ts 1003, ca 760, td 670, ca 761, td 702 , su 1011, ts 763. Note that the numbers are in octal (i.e., the digits 8 and 9 are not allowed).

Chapter 14. Physical basis for deconvolution

Aristotle: First, have a definite, clear practical ideal; a goal, an objective. Second, have the necessary means to achieve your ends; wisdom, money, materials, and methods. Third, adjust all your means to that end.

Maurice Ewing and J. Lamar Worzel

Maurice Ewing (1906-1974) was one of the towering figures in the history of geophysics. He did much fundamental work on plate tectonics. His life and achievements were brilliantly summarized in a long, three-part article that appeared in *The New Yorker* in 1974. Its beginning reads:

William Maurice Ewing was a soft-spoken, tired looking man, a Texan by birth, a physicist by training. He was 67 when he died, and had spent a large part of his life at sea, exploring the earth. He liked to be at sea for a couple of months a year, and whenever possible would fly off to join a ship in Manila or Punta Arenas. His work led him to know distant and out-of-the-way ports—their waterfronts, ship chandlers, and harbor masters— and stretches of ocean that merchant seamen never travel. He was a big man—a large, rough-looking character, he once said— over six-feet tall and broad-shouldered from hauling cases of dynamite and other geophysical equipment around the decks of ships. His face was broad, lined, and weathered, and his eyes, under puffy lids, were small and blue. He spoke of the ocean floors with as much familiarity, regret, and frustration as some other passionate traveler might speak of the antiquities of Italy. Ewing wanted to understand how the earth works. He could not wait to learn anything, wait for someone to invent a better instrument, wait until tomorrow to hear if his ships at sea had discovered something today, wait until Monday for Saturday's mail. It caused him great anguish to let an unfamiliar stretch of ocean go unexamined or a few unscheduled hours of ship time go unused, or to sail at less than full speed while his instruments recorded the characteristics of the bottom. "There are other men who think in worldwide terms," a geologist said recently, "but Ewing—sending his ships around the world over and over and over, like no one else—also did in worldwide terms."

J. Lamar Worzel (1919 - 2008), was a geophysicist known for his important contributions to underwater acoustics, underwater photography, and gravity measurements at sea. With Ewing he established that the sea floor is not a former sunken continent but an original ocean basin. Together with Ewing, he discovered "shadow zones" in the oceans that were not accessible to sonar detection as well as "deep sound channels" that transmit low frequency sounds for long distances. The Maurice Ewing and J. Lamar Worzel Professorship of Geophysics at Columbia University in New York is named in their honor.

Figure 9. J. Lamar Worzel and Maurice Ewing

Ewing and Worzel (Figure 9) took on the task of finding out about the ocean floor. The sound waves of SONAR could map the ocean bottom. Seismic waves could penetrate the sea bottom but the seismic reflection records could not be interpreted because of the interference from water reverberations. Today oil companies are finding oil below the sea floor all over the world. In 2009, oil was discovered 250 miles southeast of Houston at a depth (from the surface of the water) of more than 35,000 feet — greater than the height of Mount Everest. Such discoveries have been made possible by digital three-dimensional seismic imaging that can pinpoint where the oil and gas lie. How did we get to this point? To find out we have to go back to MIT Building 2

Ripples and waves

A wave is traveling energy. Seismic waves travel through the earth; ocean waves travel through the water; sound waves travel through the air. The closeness of MIT Building 2 to the Charles River meant that the Mathematics Department was always in sight of wave motion. Waves of one form or another can be found in an amazingly diverse range of physical applications, from the oceans to the science of sound. Put simply, a wave is a traveling disturbance. Ocean waves can travel for hundreds of miles. Seismic waves from an earthquake travel through the interior of the Earth, some all the way to the core of the Earth and making it back to the surface. Sound waves travel through the air into our ears, where we process the disturbances and interpret them as voice or music or noise.

Professor Norbert Wiener wrote the following passage in his autobiography *I Am a Mathematician*.

> It was at MIT too that my ever-growing interest in the physical aspects of mathematics began to take definite shape. The school buildings overlook the River Charles and command a never changing skyline of much beauty. The moods of the waters of the river were always delightful to watch. To me, as a mathematician and a physicist, they had another meaning as well. How could one bring to a mathematical regularity the study of the mass of ever shifting **ripples and waves**, for was not the highest destiny of mathematics the discovery of order among disorder? At one time the waves ran high, flecked with patches of foam, while at another they were barely noticeable

ripples. Sometimes the lengths of the waves were to be measured in inches, and again they might be many yards long. What descriptive language could I use that would portray these clearly visible facts without involving me in **the inextricable complexity of a complete description of the water surface**? This problem of the waves was **clearly one for averaging and statistics**, and in this way was closely related to the Lebesgue integral, which I was studying at the time. Thus, I came to see that the mathematical tool for which I was seeking was one suitable to the description of nature, and I grew ever more aware that it was within nature itself that I must seek the language and the problems of my mathematical investigations.

The above paragraph is an excellent example of how Wiener would attack a difficult problem. Wiener's vision was enormous but there was wonderful simplicity in his devices. Simplicity of device is the sign of the master. As stated above, he wanted to portray the visible facts of waves without the attendant complexity of a complete description of the water surface. He proposed that this problem of waves was clearly one for averaging and statistics.

Impulse and resulting wavelet

Many situations involve visual imagery. Visualization is especially useful in solving problems where shapes, forms, or patterns are concerned. Scientists and engineers constantly use visual thinking from initial layout to final interpretation. They think visually using an intricate sequence of mental images. They also use of conceptualization in those many instances when the desired results are not at all obvious. Imagery is found throughout the works of William Shakespeare. In the play *King Henry 6 Part 1*, the leading character is Joan of Arc. She was on the French side in war against the English. She says:

Glory is like a **circle in the water**,
Which never ceaseth to enlarge itself
Till by broad spreading it disperse to naught.

Paraphrasing Shakespeare, we may write:

> A wavelet is a circular disturbance in the water, which never
> stops expanding until it spreads so much that it disappears.

Below is a picture of the two wavelets, each caused by a raindrop.
(Actually there is a third circle, quite faint, inside the one on the right,
but for this discussion we will ignore it.) The problem is to find where
and when the two raindrops fell. The answer is given by deconvolution.
Let us do the deconvolution. The raindrops fell at the center of each
circle. We can find the location of the center from the curvature of the
circle. Because the wavelet on the right has expanded more than the
wavelet on the left, the raindrop on the right fell before the raindrop on
the left. Eventually, each wavelet disperses to zero. We have used two
raindrops. However the same analysis works no matter how many
raindrops fall.

Convolution and deconvolution

What causes such a wavelet? A puddle is an accumulation of water in a
small depression on the ground. Puddles are a source of fascination for
children. After a rainstorm, there would be a big puddle in the dirt road
beside our house. As a little boy, I dropped pebbles into the water. Look
carefully at the flat surface of still water in a puddle. Into the still water,
drop a pebble. You then will see the wavelet that Shakespeare

describes. In time the wavelet will dissipate to nothing. The pebble is the impulse that causes the wavelet. In scientific terms, the wavelet is said to be causal. It is caused by the pebble. In summary a wavelet is causal and dissipative.

Now look at a puddle in a summer shower. Each raindrop gives the impulse that causes a wavelet. Each individual wavelet spreads out and eventually dissipates. New raindrops fall to replenish the supply of wavelets. The overall result is that all of the wavelets add together to produce the ripples and waves on the puddle. The movement seen on the puddle is nothing but a large collection of wavelets, each one caused by a raindrop.

When you look at the puddle in the rain, you do not see the individual raindrops nor do you see the individual wavelets. You just see the wave motion on the puddle. What do we not know? We do not know the size, time and location of each raindrop. We do not know the shape of the wavelet. Thus there are two unknowns, namely, (1) the wavelet and (2) the impulses of the raindrops. What do we know? We know the wave motion on the puddle. We have one known and two unknown quantities. In such a case there are many possible solutions.

Known	Unknown	Unknown
wave motion	wavelet	raindrops

The raindrops are all scattered about in a rough random matter. Each raindrop is an impulse that spreads out into a wavelet. Each wavelet is an entity that can be compressed to yield the underling raindrop. The wave motion is obtained by the building up of wavelets. Whereas the raindrops form a jagged irregular pattern, the wave motion (gentle ripples and waves) is smooth and unbroken.

There is a time to build up, and a time to break down. **Convolution is a time to build up.** The wavelet convolved with the raindrops gives the wave motion. **Deconvolution is a time to break down.** The inverse wavelet convolved with wave motion the gives the raindrops. The inverse wavelet is the deconvolution operator. In other words, the deconvolution operator convolved with the wave motion gives the raindrops (the deconvolved signal).

Convolutional model

What conditions can we impose so that we can find the raindrops and the wavelet shape? Let us see. Each raindrop produces an impulse in the water at the place and time of the pebble drop. We assume that the raindrops fall at random on the puddle. Accordingly we may say that the **impulses are random**.

Each impulse causes a wavelet. The wavelet shape is as Shakespeare described. It is a circle in the water, which never ceases to enlarge itself till by broad spreading it disperses to naught. The raindrop imparts its energy to the creation of the wavelet. The wavelet has large energy at its beginning but then the energy dwindles away to nothing. Accordingly, the energy of the wavelet is concentrated at it beginning. In mathematical terms the **wavelet is minimum delay**.

The wavelets interfere with each other as they travel outward. At some places the interference is destructive in that the wavelets cancel each other out. At other places the interference is constructive in that the wavelets reinforce each other out. In mathematical language, the wave motion taking place on the puddle is the folding together of all the wavelets, each wavelet resulting from the impulse provided by a raindrop. In mathematical language, **the wave motion is the convolution of the wavelet with the impulses.**

We can observer the wave motion, so it is known. Wave motion (known) equals wavelet (unknown) convolved with impulses (unknown). We have one mathematical equation with two unknowns. How do we solve it? As it turns out, the mathematical equation can be solved for the two unknowns if we impose **two conditions; namely, the impulses are random and (2) the wavelet is minimum delay.** The physical situation indicates the validity of these two conditions. All is well! The conditions required for a mathematical solution are provided by the physical situation.

We have described how a raindrop creates a wavelet on the puddle. Let now turn to acoustics. A sound pulse is emitted from a source at time zero. The echo is the reflection of the pulse from an obstacle. The echo (reflection) arrives back at the source a certain amount of time later. This amount of time is called the echo time. Typical examples are the echo produced by the bottom of a well, by a building, or by the walls of an empty room. A true echo is a single reflection of the sound source. Let the echo time from that first well be one second. Let the echo time from that second well be two seconds. We can conclude that he second well is twice as deep as the first well. In other words, knowledge of the echo time gives the depth of the well.

Let us now turn to geophysics. A seismic pulse is emitted from a source at time zero. An echo is the reflection of the pulse from an interface between two rock layers. Call it the first interface. The echo (reflection) arrives back at the source a certain amount of time later. This amount of time is called the reflection time for the first interface. A second echo is the reflection of the pulse from the next deeper interface. Call it the second interface. The echo (reflection) arrives back at the source a certain amount of time later. This amount of time is called the reflection time for the second interface. The same can be said for deeper interfaces. The result is a sequence of impulses, where each impulse occurs at the respective reflection time. Each reflection time indicates the depth of the respective interface.

All well and good! But we have not yet considered reverberations. The use of acoustic echo and reverberation effects dates back many hundreds of years. Sacred music made use of an understanding of the complex natural echoes and their reverberation inside cathedrals. An echo chamber is a hollow enclosure used to produce reverberated sounds, usually for recording purposes. For example, the producers might wish to produce the aural illusion that a conversation is taking place in a large room or a cave.

For each seismic reflection, there is a reverberation. Each reflection is an impulse. The accompanying reverberation is the wavelet. The seismic trace is the convolution of the reflection series and the wavelet. As Isaac Newton wrote, "Truth is ever to be found in the simplicity, and not in the multiplicity and confusion of things." The convolutional model may be stated in this way. The seismogram may be visualized as the totality of responses to impulses, each impulse being associated a reflection. These responses are the seismic wavelets.

Blind deconvolution

Because the solid earth is fixed in place, seismic waves follow deterministic paths. Accordingly, from a strict point of view, there is nothing random about seismic-wave motion. Theoretically, the wave equation with appropriate initial conditions and boundary conditions can be used to describe the wave motion. However, the received seismic signal is replete with deterministic clutter (a.k.a. signal-generated-noise), mainly reverberations. The difficulties inherent in using a strictly deterministic approach would exceed computer capabilities even today. Wiener's methods are statistical. The resolution of the problem was clear. A simple statistical model of the received seismic signal was needed. What would it be? The following assumption was made. The received seismic signal is described by the general convolutional model (a.k.a. convolutional equation):

> received signal = random component (impulses) convolved with deterministic component (wavelet)

Next we are faced with the problem of specification. In the absence of any additional information, it is assumed that

(1) Each impulse is **random** (i.e., a member of a random sequence**)**.

(2) Each impulse causes a **deterministic wavelet**.

The convolutional equation involves three quantities; namely received signal, impulses, and wavelet. If two of these three quantities are known, the equation can be solved for the third. The difficultly arises when you only know the received signal and you want to solve for both the impulses and the wavelet. This case is called **blind deconvolution**. A further assumption has to be made in order to find the solution. What else can we say about the wavelet? To this end, we will now discuss the meaning of delay.

As a climber goes up a mountain, he experiences the rate of change of elevation. The gradient is the vector (since it has both magnitude and direction) that points in the steepest direction. Thus, if the climber continually follows the gradient, he will take the steepest path to the top of the mountain. Assume that the speed of the climber at any point on the mountain is the same in all directions (as in an isotropic medium). Thus, by taking the steepest path, the climber will reach the top in the shortest time. In other words, the path that follows successive gradient vectors represents the **least-time** path.

Skiing the fall line (a.k.a. flow line) means to descend on the steepest path. The fall line is in the direction of the negative gradient. The fall line will twist and turn. The skier's eyes can detect local changes in the steepness of the hill. When on the fall line, the skier feels equal pressure on both legs. To ski in any other direction would result in different pressures on the legs. Skiing the fall line means that the skier follows the direction of the negative gradient all the way down the mountain. Assume that the speed of the skier at any point on the mountain is the same in all directions (as in an isotropic medium). The skier reaches the bottom in **least time**.

Let us now state the above reasoning in terms of wave motion. The traveltime of a wave along the flow line is less than the traveltime along any other path. In other words, the flow line is the least time path. Pierre de Fermat (1601-1665) formulated this rule, which is known as Fermat's principle of least time. In terms of Parkinson's Law of Delay, **traveling waves do not procrastinate**; there is no extra delay.

The additional assumption required for blind deconvolution is: the wavelet is **minimum delay**. In such a case, blind deconvolution can be done. This whole book is about blind deconvolution. For brevity we omit the word "blind" and use only "deconvolution" throughout.

Waves are disturbances on the ocean in the form of moving ridges or swells. A moving ridge is called a wavefront. At the seashore, you can watch each wavefront travel until it breaks upon the shore. Shakespeare's *Sonnet 60* reads:

> Like as the waves make towards the pebbled shore,
> So do our minutes hasten to their end;
> Each changing place with that which goes before,
> In sequent toil all forwards do contend.

A charge of dynamite on the surface of the ground initiates a downgoing seismic wave. The deep subsurface interfaces produce reflected seismic waves. In succession, the reflected waves make their way toward the surface of the earth. The geophones detect these waves which are recorded as traces (received signals). The reflected seismic waves are like the water waves in Shakespeare's passage.

Each reflecting interface produces an **impulse**. The accompanying reverberation represents the **wavelet**. The seismic trace is the **convolution** of the impulse sequence and the wavelet. The problem of deconvolution is to decompose the trace into its two components, namely the sequence of impulses and the wavelet. One quantity (the trace) is given, and two quantities (the impulse sequence and the wavelet) are unknown. As a result two additional conditions must be added. The first condition is that the impulses are **random**. It rests upon insufficient knowledge. The second condition is that the wavelet is **minimum delay**. It rests upon least time (i.e. no procrastination).

The seismic trace with all its ripples and undulations is known. From this information, the deconvolution operator is computed. From the mathematics it follows that the deconvolution operator is the inverse of the minimum-delay wavelet. Consequently the deconvolution operator can be used to remove the wavelets from the seismic trace. The result is that we are left with the impulses. The impulses give the desired reflection times, from which we can obtain the depths of the underground interfaces. As an extra benefit we can invert the deconvolution operator to obtain the minimum-delay wavelet.

The wavelet

Dean Clark (*The Leading Edge*, volume 4, 1985) writes

> "The oil company people were generally very smart and they asked the right questions. Asking the right questions is frequently half the battle," Robinson says. "It was they who kept getting us back on the track when we'd get too far ahead of ourselves. In 1953 we were trying things that were just too complicated. Some of the multi-trace things we were attempting then haven't yet been completely explained. The message I got from industry was, 'Slow down. Get something simple and dirty but which works.' That saved the day. I went back to working with a single trace and that's what led to my thesis. Otherwise, I'd have been going in that other direction indefinitely."

It was clear that in 1953 the oil company geophysicists were not particularly interested in learning about the complexities of Wiener's mathematics. The idiom "as easy as pie" is used to describe a pleasurable and simple task. The idiom refers to eating a piece of apple pie. The first attempt of the GAG was to make, **not the theory** deconvolution, but **instead the computation** of deconvolution as easy as pie. The digital computer achieved this goal. In 1953, Raytheon was ready to perform seismic deconvolution. The oil companies could outsource their seismic records to Raytheon. However, at that time, the oil company geophysicists, for the most part, showed signs of computer anxiety. They liked the way that they were doing things. To abandon their familiar analog methods and venture forth into the unknown domain of digital techniques did not appeal to them. They did not take advantage of the digital computer service that Raytheon offered. However, the Raytheon involvement did spur the oil companies to change the recording of seismic traces from photographic paper to magnetic tape. One could amplify a trace (i.e., multiply the trace by a constant value), shift a trace in time, add traces, and display traces in various formats. With the use of magnetic tapes, the oil companies developed analog processing methods that were important strides forward.

In research the GAG was exploring the complexities of multichannel deconvolution. I was obsessed with the problem of locating hidden reflections. That was achieved by the deconvolution. The GAG and Raytheon had computed hundreds of deconvolved seismic traces. The hidden reflections were revealed. The discussion in 1953 was: "The world is analog, so why not deconvolution? Find an analog way to do deconvolution and then we will consider it. " Ten years earlier in January 1943 Norbert Wiener's Section D-2 contract was not renewed. His work on prediction was largely unappreciated in an analog setting. Nothing had changed; analog was still supreme.

Frederic Chopin (1810-1849) was a Polish composer and a virtuoso pianist of the Romantic era. His genius was "without equal in his generation." Chopin wrote:

> Simplicity is the final achievement. After one has played a vast quantity of notes and more notes, it is simplicity that emerges as the crowning reward of art.

The GAG computations "played a vast quantity" of deconvolved seismic traces. But what was the "simplicity that emerges as the crowning reward of art?" And then the answer came! **The "simplicity" is the wavelet.** The wavelet is the final achievement. More specifically, the existence of a physical seismic minimum-delay wavelet was affirmed by computation. From the time of Pythagoras until then wavelets were either actually seen on the water or mentally seen as mathematical constructs. No wavelet had ever before been computed from observational data. The physical wavelet (shown in Figure 10.) computed from seismic data was the highpoint in my PhD thesis in 1954. Geophysicists took up the digital wavelet in research and have produced remarkable results. More papers are written on wavelets today than those on deconvolution. In the early 1980s Jean P. Morlet, a French geophysicist, applied wavelet knowledge to devise the cycle-octave transform. In 1984 Morlet and Alexandre Grossman developed the rigorous basis of the cycle-octave transform and it became known as the wavelet transform.

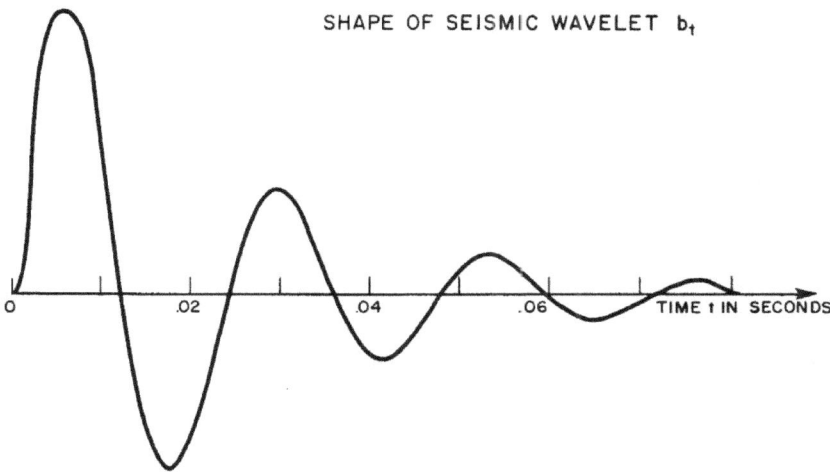

SHAPE OF SEISMIC WAVELET b_t

Figure 10. Depiction of the first physical wavelet ever computed from observational data. Seeing is believing.

The word "wavelet" goes back to Christiaan Huygens in 1690. There is a traditional story that as a boy Huygens watched the water on the canals in Holland near his house. Wavefronts were running before his eyes. How were successive wavefronts related? In answer, Huygens reasoned that a wavefront is formed by all the wavelets emanating from the previous wavefront. Huygens Principle appears in every elementary physics book. It states that light propagates as waves and every point on a wavefront is a source of spherical wavelets. The wavefront at any instant is the envelope of spherical wavelets emanating from the wavefront at the prior instant.

For simplicity, consider the case of a homogeneous isotropic medium, so all the secondary spherical wavelets have the same radius. In such a case, the medium represents a linear space-invariant system. In engineering terms, the wavefront at the initial time is the input to the system, the wavelet is the impulse response of the system, and the wavefront at some later time is the output of the system. Thus Huygens principle, in fact, states that the wavefront at some later time (i.e. the output) is equal to the convolution of the wavefront at the initial time (i.e. the input) with the wavelet (i.e. the impulse response). In other words, Huygens principle is a spatial convolutional model in which the wavelet plays the key role. Despite the prominence of this principle, the Huygens wavelet always remained as a conceptual concept in the historical development of physics. Two centuries had to pass before Huygens concept of wavelet was given mathematical form. Green's function is the solution of the wave equation with an impulsive source function. The Green's function is the wavelet.

My 1954 Ph.D. thesis at MIT introduced the convolutional model. In exploration seismology, a source of energy is initiated on the surface and the resulting seismic waves travel down through the earth and are reflected from various interfaces and boundaries in the subsurface layers. The returning waves come back to the surface where they are recorded as a seismic trace. The basic problem is that there is a host of other signals coming in, such as multiple reflections, reverberations, diffractions, surface waves, refracted waves, all of which hide the desired primary reflections. The problem is how to unscramble this maze.

For example, in seismic exploration at sea the water layer acts as an imperfect lens that masks the reflections from depth. In other words, reverberation in the water layer hides the signals coming from the geologic interfaces. The reverberation, which is a slowly damped oscillation, represents the wavelet. For such a case, the **convolutional model** states the recorded seismic trace is the convolution of the reverberation wavelet with the reflected signals coming from depth. Having the convolutional model of the trace, the next step is to deconvolve the trace. In other words, unscramble the trace to get back the components. The deconvolution operator is the inverse of the seismic wavelet. Deconvolution removes the wavelet, thereby making the recorded trace reverberation-free.

The **convolutional model** would be the basis used for making Wiener's mathematics as easy as pie. The two physical quantities are (1) the impulses (e.g., the reflectivity of the underground interfaces) and (2) the wavelet (e.g., the seismic wavelet). The convolution model is: The seismic trace is the convolution of these two quantities.

The Raytheon computations in 1953 gave three hundred examples in which the prediction errors indicated the reflections of the underground layers. That gave a linkage of the mathematical random sequence with the physical seismic reflectivity. Accordingly there must be a physical linkage of the mathematical minimum-delay wavelet with a physical seismic minimum-delay wavelet. In other words, to the mathematical convolutional model there was a physical convolutional model as counterpart. It is the connection of the physics with the mathematics that matters. In such a case, the mathematical model of a physical system can be used to extrapolate further properties of the physical system in question.

Professor Tadeusz Ulrych

Tadeusz (Tad) Ulrych was born in Warsaw, Poland, and fled the country with his parents following the Nazi invasion in 1939. He obtained a BSc degree in electrical engineering at London University, United Kingdom. He moved to Canada and received his PhD degree (1963) at the University of British Columbia (UBC). Following postdoctoral fellowships at Oxford University and at the University of Witwatersrand, South Africa, Ulrych joined UBC, where he was a professor of geophysics until he retired in 2000. He has made many outstanding contributions to exploration geophysics and was the recipient of numerous awards. He wrote the article "The whiteness hypothesis: Reflectivity, inversion, chaos, and Enders," published in GEOPHYSICS, vol. 64, No. 5 (1999); p. 1512-1523. The non-mathematical portions follow:

ABSTRACT. The earth's reflectivity is one of the principal targets of seismic exploration for oil and gas. It exhibits some fascinating properties. It is blue. It is fractal. In all probability, it reflects the output of a chaotic generator. I examine some of these characteristics here. Specifically, I look at the statistical properties of a canonical chaotic model, and I examine the effect of multiples on the statistical properties of seismograms generated from primary reflectivities. This article is a tribute to Enders Robinson. He is a pioneer who has inspired many. He inspires many to this day. I am one of them.

INTRODUCTION ...   I began my upgraded education then and continue it to this day. It was, in my opinion, to a large extent the work of Enders Robinson that transformed applied seismology and made the study what it is today, profound and at the edge. I will describe this work briefly because it has made such impact and because it illustrates so clearly the points I want to emphasize.

Before I do this, however, let me talk briefly concerning the title of this article. Whiteness because that is at the core, as are reflectivity and inversion. Enders because, to such a great extent, he started it all for me (and for us). Why chaos? Is it here just to add some pizzazz, a catchy word that attracts? Not at all. I believe very strongly that chaos is a paradigm for a philosophical (and mathematical) understanding of our earth (and music and architecture). It gives one a manner to rationalize determinism with unpredictability. For those who believe in an Almighty, to put into place the deterministic hand of God with the stochastic outcome of free will. The philosophical aspect is very important to me. As time passes, one ages. With age, one becomes more consciously philosophical. I believe that one does so simply because one can no longer cope with the detail of it all (nor actually desire to). We leave the detail to those who can cope with it—the young. With the passing of time, we begin to assimilate and synthesize, to philosophize. It is not without reason that most of the philosophers whom we admire and respect are dead.

THE IDEAS OF ENDERS ROBINSON. The model for the seismic trace is ultimately simple (a property that is encouraging to some and discouraging to others). It is linear and, as such, embodies the principle of superposition. ... Because of linearity and time invariance, the resulting seismic trace is the convolution of the wavelet and the reflectivity sequence.

It is the reflectivity sequence that is the goal of the interpreter. To achieve this goal, the wavelet must be removed. And here is the impasse. We do not, in general, know the wavelet, and the convolution equation is an equation with two unknowns. Enders Robinson (1967 publication in *Geophysics* of entire 1954 PhD thesis) was, I believe, the first person to seriously tackle this ill-posed inverse problem in seismology. He approached it in the only way that makes sense. By imposing a priori information—information based on physical and geological insight. I stress this point here. Enders approached the task of finding a plausible solution from the infinity of possible solutions—not by an ad hoc regularization such as ridge regression, but by carefully reasoned physically motivated constraints. He required and proposed two solutions.

The primary reflectivity of the earth is white.—White means that all frequencies exist and are represented with equal power. Consequently, the autocorrelation is a delta function at zero lag and the process is completely uncorrelated. This is an absolutely reasonable hypothesis. Surely the sedimentary generator is not aware on Monday what layer it will deposit on Tuesday. The reflectivity sequence must exhibit no memory and certainly has no predictable features.

The seismic wavelet is minimum phase.—(Or minimum delay; it's the same thing.) What does this actually mean? I believe, the physical meaning of this constraint is an expression of Fermat's principle. The elastic energy that travels through the earth as a result of an impulse on the surface arrives at the receiver in the minimum time possible. The energy of the arriving pulse therefore tends to be front loaded. Robinson expresses this phenomenon in his minimum delay theorem (Robinson and Treitel, 1980). This hypothesis feels correct. Given a delta function at the source, the wavelet would indeed be minimum phase. The problem is that the initial energy input is not an impulse but, in all likelihood, some non-minimum phase series of impulses that represents the oscillation of the cavity following the detonation of the physical charge. For this reason, marine sources are much more likely to give rise to minimum-phase wavelets.

... [Mathematical and scientific part of article is omitted.] ...

ACKNOWLEDGMENT. This article is meant as a tribute to Enders Robinson. His footprints are discernible on most of the scientific beaches that I have visited. He has been and remains my teacher. He has and continues to inspire my work. Thank you, Enders.

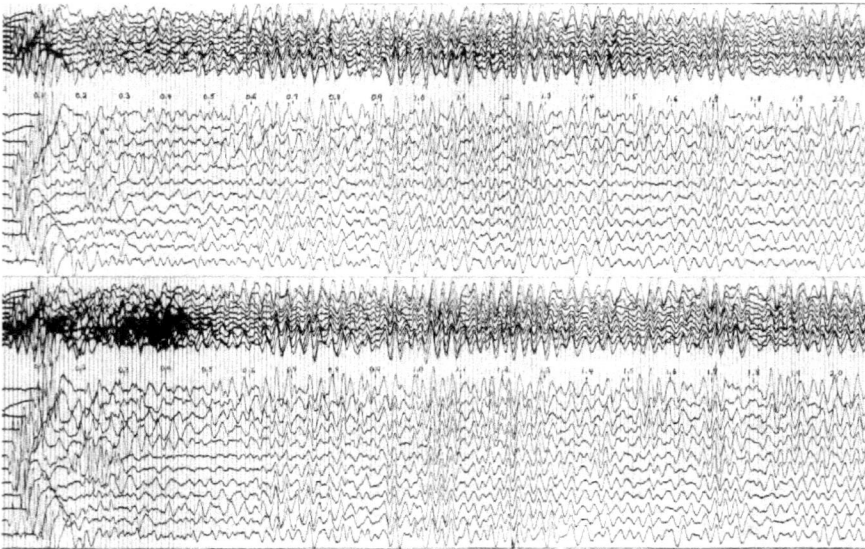

Chapter 15. Aftermath

Max Planck: "A new scientific truth does not triumph by convincing its opponents and making them see the light, but rather its opponents eventually die, and a new generation grows up that is familiar with it."

Proposal by Raytheon in 1954

In 1947 Dr. Richard Clippinger had worked with John von Neumann to redesign the ENIAC. The result was that the ENIAC was transformed into the world's first stored-program digital computer. Now in 1954 Clippinger was ready to do the same for geophysical exploration. Clippinger and his team were ready to transform exploration into the world's first industry to use digital signal processing. Raytheon would obtain or build all the elements required from input to output. Raytheon would use digital magnetic tape and build the necessary analog/digital and digital/analog converters. Raytheon would either obtain or build digital computers suitable for seismic work. In particular, Raytheon proposed using the new IBM 701 computer. Raytheon had the great advantage of understanding the mathematics involved, of having written the necessary computer codes, and of successfully performing digital signal processing on oil company seism c records. Raytheon had a close connection with MIT, especially with the radar technology being developed at MIT Lincoln Laboratory.

On March 29, 1954 Raytheon submitted the following proposal to the oil industry.

UTILIZATION OF ELECTRONIC DIGITAL COMPUTERS IN ANALYSIS OF SEISMOGRAMS

R. F. Clippinger, B. Dimsdale, J. H. Levin
Computing Services Section, Raytheon Manufacturing Company
Waltham, Massachusetts

1. INTRODUCTION.

This report is supplementary to MIT GAG Report No. 6, and a sequel to our previous report which was included as Section 7 of MIT GAG Report No. 5.

1.1 Role of Raytheon. Certain programming, coding, and computational tasks relating to the MIT study on the analysis of seismograms were subcontracted to the Computing Services Section of the Raytheon Manufacturing Company. More specifically, Raytheon's responsibilities were:

(1) to program for a digital computer the processes of

(a) fitting of a linear prediction operator to multiple time series such as those occurring in seismic recordings;

(b) predicting by means of this operator, and computing running averages of the squared errors of prediction;

(2) to apply the code obtained in (1) in about 350 cases.

The code was prepared for the Ferranti computer (FERUT) at the University of Toronto, and the results and conclusions are presented in MIT GAG Reports Nos. 4, 5, and 6.

1.2 Raytheon's Computing Services Section. The Computing Services Section is an outgrowth of Raytheon's activity in the design and production of computers and computer components. The activities of the Section are directed toward fulfilling computing needs arising in private industry, the government, and Raytheon itself. The Section provides the services of mathematical analysis, programming, and coding for electronic digital computers. Problems are programmed for the equipment best suited. The Section has arrangements with leading computation laboratories for use of their equipment (notably with the IBM Corporation for use of its IBM-701, with Remington Rand, Inc. for use of the UNIVAC, and with the University of Toronto for use of the FERUT). In addition it has its own small scale computer, the Raytheon SC-101, a machine with a 4000 word drum memory, and operating at about 10 operations per second. A larger computer is to be acquired.

Additional services include consultation and assistance in the evaluation and selection of computers for specified purposes, The scope of the Section's activities includes problems in the fields of engineering, statistics, hydroelectric power generation, flood control, accounting, and other bus ness applications, to name but a few of the areas where work has been carried on. The staff now consists of about 16 persons - double its size at the time of our previous report - representing a total accumulation of some 50 years' experience in all phases of numerical mathematical analysis and digital computer applications.

2. CURRENT VS. POTENTIAL MECHANIZATIONS OF SEISMIC ANALYSIS ON DIGITAL COMPUTERS.

Our activity on the subject project has resulted in a code for the FERUT for analyzing seismograms on the basis of the statistical approach described in MIT GAG Nos. 5 and 6. This code was designed to handle all the cases arising in the study. It provides for the construction of operators on the basis of up to 4 traces, 4 lags, and operator time intervals of up to 60 readings. (Actually the program is more flexible than may appear from this description. By a simple taping mod fication one can take 16 lags on one trace, 12 lags on one trace, and 4 lags on a second, 8 lags from each of two traces, 8 lags on one trace and 4 lags on each of two other traces, or 4 lags on each of 4 traces.) The computer output consists of the operator coefficients and a tabulation of running averages of squared errors over the predicted range.

The MIT study indicates that the power of this technique can be increased by using up to 20 or so lags, and up to a couple of hundred readings in the operator time interval.

Our own studies further suggest a schedule of activity such as the following leading towards the practical analysis of seismograms on a quantity scale:

(1) coding the problem for larger, faster and more versatile digital computing equipment now available to us;

(2) adapt the program so as to incorporate the MIT conclusions noted above;

(3) adapt the program further so as to substantially reduce both the clerical time spent in data punching and handling, and the supervisory time, by passing many of these functions over to the computer;

(4) make use of now commercially available analog to digital conversion devices to simplify and speed up reading of seismogram traces and punching of data;

(5) likewise make use of commercially available digital to analog conversion equipment to simplify and speed up drawing of error curves from computer output.

For cases similar to those we have already done, a program such as the above would have the triple effect of

(1) reducing cost of turning out error curves by 60 percent or more;

(2) stepping up overall speed by a factor as high as 10;

(3) increasing the power, flexibility, and range of applicability of the current code.

Within such a framework we can readily meet the individual needs of prospective clients. For example the code resulting from the plan outlined above would equally well permit the determination of any of the 6 operator types defined in Section 1 of MIT GAG No. 5. Or it allows for individual preferences in the measure of effectiveness of prediction to be used (Section 5, MIT GAG No. 5).

In view of the experience we have thus far gained on this type of problem, we estimate that we could complete a program accomplishing the foregoing aims in a matter of 2 to 3 months. We can in the meantime, of course, render these services by the use of our present code for FERUT.

We can, further, depending on the desires of the individual client, code such additional features as weathering corrections, and automatic selection and recording of reflection times. Or, as an alternative to the statistical approach, we could program the linear operator discrete filtering technique [cf. Introduction, MIT GAG No. 6) .

In the longer run we can look forward to the prospect of feeding original magnetic tape recordings of seismic data into a digital computer. Equipment has already been successfully designed for converting magnetic tape data from analog form, such as those produced in seismic recordings, to magnetic tape data in digital form for purposes of feeding into a digital computer.

3. SOME ALTERNATIVE WAYS OF PROCEEDING WITH A COMPUTATION PROGRAM.

There are three principal ways for firms engaged in oil exploration to proceed with the above kind of program. It is possible to

(1) acquire one's own computer by purchase or rental;

(2) buy time on someone else's computer, doing one's own programming and coding;

(3) make use of a commercial computing service organization turning over all programming, coding, and operation.

The choice of the most economical of these ways depends on individual circumstances. The writers of this report have collaborated in an intensive discussion of this matter in a series of articles copies of which are attached. We shall summarize a few main points here and attempt to crystallize the ideas by means of an example.

Today's electronic digital computers vary greatly' in speed, storage facilities, coding systems, checking facilities, input-output methods, and overall design. In price they range from about $ 50,000 to $1,000,000; on a time purchase basis (such as in (2) above) from about $30 to $300 per hour; and on a rental basis somewhat less than this (the exact amounts depending on how machine costs are written off, the allowances for overhead, number of hours per week that equipment is operable, etc).

For this range of prices, or hourly rates, computing speeds range from 10 to 6000 operations per second, internal memories from 10,000 to 600,000 binary digits (or equivalent), auxiliary storages up to 100,000,000 binary digits; input-output speeds up to 10,000 characters per second. Costs per operation will be found to decrease as speed and cost of computers increase, ranging from about 1000

What does this array of possibilities mean as far as analyzing seismograms is concerned? We shall see that using the most inexpensive computer is not necessarily the most inexpensive way of doing the job. Let us, for definiteness, consider the computer costs only for inverting a 20th order symmetric positive definite matrix (Section 6, MIT GAG No. 5) whose elements consist of 5 decimal digit numbers. Such a routine is an integral part of the seismic analysis procedure. We shall consider the costs on a large, a medium sized, and a small computer having the following features respectively:

Computer	Hourly Rates	Computing Speed (operations/sec)
Large	$300	6000
Medium	$100	300
Small	$40	20

Estimated times and costs for this operation on the three computers selected are given in the following table:

Computer	Time (min.)	Rate per min.	Computation cost
Large	0.2	$ 5.00	$1.00
Medium	4.0	$1. 67	$6.68
Small	60.0	$0.67	$40. 20

It should be kept in mind, that the above costs do not by any means represent the total cost of doing the problem. They do not include for example, programming and coding costs, the cost of having originally checked the program in on the computer, nor the cost of reading the program into the computer. In the case of seismic analysis, we would have to consider further the costs of measuring of traces, punching of data, drawing of error curves, and miscellaneous handling. We may also have to consider the programming difficulties that may arise because of the limitations of small computers, thus increasing the problem preparation costs in such cases.

The example above clearly indicates the greater efficiency of the larger computer. As far as the prospective user is concerned, the advisability of acquiring the larger computer depends to a great extent on the total volume of work of all kinds to be done. However, through alternatives (2) and (3) an organization is able to enjoy the economies of the largest scale computers even though the workload does not justify the acquisition of one.

It is an inadequately recognized fact that mathematical analysis, programming, and coding costs may easily constitute a much more formidable item than computer costs, This fact is generally overlooked in newspaper and other journalistic publicity, with the result that. many potential users of computing equipment have totally unrealistic ideas concerning digital computation costs. In the attached series of articles we have tried to set the record straight by explaining what kinds of talents, training, abilities, and responsibilities enter into preparatory costs, and the nature of the effort required to get a computer code into operation.

Actually, the solving of a problem on a digital computer is usually the culmination of very extensive numerical analysis, programming, coding, and other activities ordinarily not considered newsworthy. In a large computing installation which handles an impressive variety and large volume of work, the staff costs (including overhead) for problem analysis and program preparation could well run over $ 1,000,000 per year. It is a simple matter for a problem requiring several man years of preparatory effort to be run off on a computer in a matter of a few hours. Of course this same problem may have required several thousand man years of work on a conventional desk calculator. Preparatory costs for a code represent an initial investment which may be spread over subsequent applications. In the case of seismic analysis, we have a favorable situation insofar as the number of applications would be very large. Moreover, the fact that we have now had extensive experience on this kind of problem, that we have a working code for the FERUT, that we have since had extensive experience on new equipment better suited to this problem, means - so far as the more ambitious program we have outlined is concerned - getting into operation faster (probably 2 to 3 months), incorporating the desirable, cost saving features in the program which we have described, and giving prompt, reliable, service using the best equipment now available for this task.

REFERENCES

1. Automatic Digital Computers in Industrial Research, I, II, III, R.
F. Clippinger, B. Dimsdale, J. H. Levin, Journal of the Society for
Industrial and Applied Mathematics, September and December,
1953 and March, 1954.

The proposal by Raytheon gave each company the opportunity to
experiment with deconvolution apart from the other oil companies. Oil
companies were familiar with such arrangements. For example, several
oil companies would use the services of the same geophysical company,
knowing that all secrets would be honored.

You can lead a horse to water but you cannot make it drink. In other
words, you could lead the oil companies to Raytheon but you could not
make them drink. The oil and geophysical company representatives
were not ready to undertake digital seismic processing at this time.
They were discouraged because an excursion into digital seismic
processing would require new effort, and still it might fail because of
the unreliability of the existing computers. The oil companies harbored
considerable fears associated with technological change from analog to
digital. They were all trained in analog. Digital was new, different and
largely untested. They were not yet ready to convert from analog
methods to digital methods.

Geophysics was an established branch of classical physics. Geophysical
research at the time involved such things as the theoretical solution the
elastic equations of motion. Wiener's theory worked directly with
unwieldy data, and still worse, it made non-deterministic (i.e.,
probabilistic) assumptions in order to average out unwanted things.
Worst of all, digital signal processing was completely dependent on a
digital computer, which was an unknown quantity to most people.

Digital computer in 1954

My PhD thesis was reproduced as MIT GAG Report No. 7, *Predictive
decomposition of time series with applications to seismic exploration*,
dated July 12, 1954. Not only did the prediction errors give the seismic
reflections, but there is a minimum-delay seismic wavelet that

represents the physics of the situation. Today the empirical wavelet is found not only in geophysics everywhere in science and signal processing. Part of the section "Summary as given by R. R. Shrock" in Chapter 1 is repeated here.

> The theoretical work of the GAG involved the linkage of the mathematical work of Professor Norbert Wiener with the physical theory of seismic wave propagation: namely, that the **phase-characteristic** inherent in the design of a Wiener digital filter is in fact the same phase-characteristic that the earth imposes on a propagating seismic wave. The empirical work of the MIT GAG involved showing that these digital methods could automatically transform, within the computer, a field record that could not be interpreted by existing methods into a record that would yield the required information about the subsurface structure. [Added note: The term "phase-characteristic" refers more specifically to minimum phase characteristic. The term "minimum phase" has the same meaning as "minimum delay."]

In the early 1950s, exploration geophysics was a large enterprise with an analog way of doing things. It was a hit-or-miss operation with a poor success ratio. MIT did its best to convince the exploration industry to convert to digital geophysics, but was unable to make the old-timers to see the need. However MIT and MIT publications fostered a new generation of geophysics students both at MIT and elsewhere who were conversant in the art of the digital geophysics. It was they who went forth and executed the conversion from analog to digital. The conversion was complete by the late 1960s. In the 50 years from 1966 to the present year 2016 about one trillion barrels of oil have been discovered by digital geophysics. The same holds for the tremendous amount of natural gas also discovered. Yet these discoveries are but the tip of the iceberg. This amount of oil and gas could never have been discovered without the digital computer. Coal is fossil plant life and oil is fossil animal life. Such a great abundance of fossil life attests to the astonishing fecundity of the earth. Oil can be used either as fossil fuel or as fossil food. Bacteria eat oil and the bacteria go up the food chain to become food for fish. Fossil life becomes living life to complete a great feedback loop.

In mythology, Prometheus was a Titan who surpassed the whole universe in mechanical skill. Prometheus created the first man out of clay. Prometheus stole a portion of celestial fire from the chariot of the sun and animated his creation with it. The man moved, and immediately thought and spoke. Prometheus' theft of celestial fire made Jupiter very angry. To punish humanity for the stolen secret of fire, Jupiter ordered Vulcan to make a woman of clay. Her name was Pandora, and she was more accomplished than the man. Jupiter gave Pandora a sealed box with orders not to open it. However, the box was opened. Out flew every kind of disease, sickness, calamity and evil imaginable. Hope only remained inside the box. From that fatal moment, misfortune and trouble have always afflicted humankind.

The digital computer is created from clay (silicon). Will Prometheus steal a portion of celestial fire and animate the computer with artificial intelligence? If the computer were animated, would there be a sealed box with orders not to open it. Would Pandora's Box be opened? We have treated the natural resources of the Earth as if they were ours for the taking. Instead we must treasure them and husband them. We must treat the air, the water, and the land with respect. They are precious resources. We need fresh air, clean water and fertile land. We must respect nuclear wastes, and properly store them. The planet Earth is our home and we must care for it and all of its inhabitants, both plant and animal. We must live within our means. We must and can meet this challenge.

Situation in 1954

All of the analog equipment in an oil company laboratory could be replaced by one high-speed, stored-program electronic digital computer. Then deconvolution and everything else (the corrections, gain control, adjustments, mixing, filtering, correlation and spectral analysis, etc.) could be done under program control by this one computer. But the acceptance of digital seismic processing was not to be, at least not in 1953 and 1954. The oil and geophysical companies were pleased with the deconvolution results but were disheartened by the unreliability of the current state of digital technology. The consensus was that the GAG should find analog ways to do deconvolution.

In 1943 the MIT servomechanism laboratory did not renew Wiener's contract because they could not implement Wiener's mathematics on an analog computer. The GAG showed that it was simple to implement Wiener's mathematics on a digital computer. With misgivings about the digital computer, the oil companies now wanted us to implement Wieners mathematics on an analog computer. In 1953 the GAG faced exactly the problem faced by Wiener in ten years earlier. To everything there is a season, turn, turn, turn.

Computers were in their infancy. The available digital computers were not entirely suitable for geophysical processing. However, each year from 1946 to 1953, there had been a constant stream of improvements in computers and this development was accelerating each year. Oil exploration is a serious business. With patience and with time, the oil and geophysics companies would carry out the conversion to digital processing. It would happen when the need for hard-to-find oil was great enough to justify the investment necessary to turn no-reflection seismograms into meaningful data.

The oil companies had their own plan of attack. Seismic wave motion was recorded as traces on paper. The recorded wave motion could not be played back. Edison invented the recording of wave motion on a disk. The recorded wave motion could be played back. Instead of a disk, the oil companies used magnetic tape. With magnetic tape, the time origin of the traces can be shifted with respect to each other according to some scheme. The shifted traces can then be added up to give a composite trace. The composite trace is called the stacked trace. The common-depth-point stack (CDP stack, horizontal stack) made use of traces that correspond to the same common depth point but which originate from different seismic profiles and different offsets. The technique reduces the amplitude of incoherent noise, multiple reflections, and diffractions. Magnetic tape was the analog computer of the 1950s. However, all that manipulation could be done so much more easily by a digital computer. Raytheon offered this digital solution, but the oil companies did not understand.

Berthing of supertankers

During the 1940s, supertankers could be 500 feet long with capacity of 16,000 DWT. Today supertankers are as long as 1,300 feet (one-fourth of a mile or 400 meters long) with capacity of 500,000 DWT. The berthing of a modern supertanker is an art. It involves combinations of variables so numerous and complex that no amount of detailed predetermined instruction can bring a ship through a canal or dock it. The exact influences acting on a moving ship are different each time. The ship responds to each one of them. Many industrial procedures involve definite actions at specific times, and they can be routinely performed. Ship-handling is not one of them. It is the job of the pilot to land a ship.

Eastern Canada has a cold climate and oil is needed to heat homes and power industry and transportation. The terminal at Point Tupper in Nova Scotia can accommodate supertankers, but the landing involves a tricky passage up a narrow channel. Here are some comments of a pilot.

> From the time the tankers leave the Persian Gulf, we begin to hear that the ship is due at such a time. Personally I've never seen one arrive according to the computer's estimation yet. A year ago last February I went out for a Norwegian tanker of 230-some-thousand tons and we waited hours and no sign of her. Came back in and then got word she was delayed and the master was so concerned he forgot to notify the pilots. Next morning went out and got him and the tanker was encased in ice on one side of her—ice a foot thick—and it took about two hours work coming up the bay there to free the windlass and winches to be able to work lines when you got into the dock. The master told me as she came up over the continental shelf, a

tremendous sea running—the seas were coming right aboard, he had to slow her down. Talk about those tankers are like floating islands. They're not floating islands with a sea raging and solid water over the bow and spray further back turning to ice and adding greatly to the weight. They chopped an area in the ice by the ship's rail wide enough for me to get aboard. But the tanker wouldn't completely encase in ice because that oil is kept at a temperature of 135 to 140 degrees Fahrenheit. It might be freezing on the deck and railing, but the ship is a heating element in herself.

You ask the master the ship's handling characteristics—and you determine this yourself. A ship coming in when you board her is on a course of 270 degrees. Well, the first alteration may be twenty degrees working up toward the buoyed channel–270 to 290 or so—and you have three miles there to see how the ship will handle steering. And of course you have that straight run up the buoyed channel. Then we have a fifty-two-degree turn to port. That is where you really get the feel of what the ship will do—and that is where I make sure—not attempt to at all—I make sure—what I consider an element of safety—that when I'm arriving at the area where I'm beginning to make this turn that ship is not going any more than four knots. Then if the ship should be sluggish on making the turn you can increase the revolutions. You can increase the revolutions of the engine and yet not increase the speed noticeably, because it's only for a matter of two or three minutes turn the ship and, as the rudder is put the opposite direction, steady her up. The thing is, going slow, if that ship has a breakdown in her steering equipment you ring Full Astern and you can stop the ship and hold her. And you have the tugs alongside—the tugs are alongside before we make that dogleg turn on the big ships—but if you were doing six or seven knots I wouldn't guarantee what would happen. They might handle just as well but I have no intention as long as I am on the job of ever finding out. I have used the tugs on the turn but that was just to see what they would do—they were

new—but now I never use the tugs on the turn. I took a tanker
in a couple of weeks ago, she was 1141 feet long—made every
turn herself.

Let us make an analogy. In 1954 analog exploration was comparable to
a tanker of 16,000 DWT. But all of a sudden digital exploration appears.
It corresponds to a tanker of 500,000 DWT. Digital exploration is in the
offing encased in ice on one side of her—ice a foot thick. A pilot is
needed to berth the digital ship. In 1954 the exploration industry was
not ready. It has no pilots. Digital exploration would have to remain
encased in ice. However, in the 1950s MIT and other universities began
training students in computers so they could dock the great digital ships
that were to come. By the 1960s this effort succeeded and has ever
since.

Digitization of Western Geophysical

It will be remembered that in 1953 Henry Salvatori as head of Western
Geophysical had no interest at all in digital. However eleven years later,
in 1964, Salvatori had changed his mind. He appointed Dr. Carl Savit as
"director of computing" at Western Geophysical. Savit first carefully
prepared for the transition from analog to digital. On June 1, 1964
Salvatori told Savit to begin the digitization process. By the end of the
week, Savit had spent $2.5 million to get things started.

Savit found that experienced software programmers were very good
but very expensive. Unfortunately Salvatori held Savit tight to the
budget for the software programmers. Salvatori was beside himself that
computer programmers could be so expensive. Salvatori had put
together some of the original analog equipment back in the early 1930s.
Salvatori was still totally unfamiliar with digital equipment. He used to
say: "Just get the machine and then you can get ready to use it." Savit
pointed out that that's not the way it works. Western was already
behind in software. Finally, Salvatori let Savit have his way. Savit paid
about twice as much for a top-notch software expert as for an
experienced engineer.

Salvatori still believed in analog devices, as most of the old-timers did. One time, in the early stages, Salvatori showed Savit a seismic record and asked him to change some amplitude. Savit said he would get back to Salvatori in about four hours. Salvatori was astonished. He said, "With all this fancy digital stuff? This is only a matter of turning a knob on an analog system!" Savit shrugged and said that things had changed.

Salvatori was nervous and apprehensive about anything digital. Nothing seemed to be happening. In September Salvatori retained a consulting company. The company sent an expert to interview the people at Western Geophysical. The verdict of the expert was that Western was doing everything wrong. Western was advised to cancel all computer orders. The consulting company would then form a team of experts to study the whole problem again. Such a study would take about 12 to 18 months, after which Western could reinstate orders with assurance that all would be optimal.

Deconvolution had open up immense new areas to exploration, including Alaska and most water covered regions of the word such as the North Sea and the Gulf of Mexico. A delay would cause Western to lose a lot of business. Western understandably decided to go on with the original plan. In May 1965, Western turned out its first deconvolved seismic section. In so doing, Western had significant help from IBM. The software programs for digital processing were totally different from anything IBM had ever been done before.

Salvatori never fully accepted that digital processing had achieved what geophysicists had desired for years. Salvatori was not overly impressed by the processing capabilities of the whole scheme. Salvatori was not alone in having difficulty adjusting to digital. Savit said that many geophysicists simply did not accept the change and retired or retreated into the woodwork, as it were, so as not to have to deal with it.

Acceptance of digital exploration in 1964-1966

In 1954 no one in geophysics wanted to embrace digital technology. The thinking was strictly analog. At that time geophysicists had a large investment in analog techniques, as did everyone else in the science and engineering. Thomas Edison said, "Our greatest weakness lies in giving up. The most certain way to succeed is always to try just one more time." The GAG deconvolution programs were working. The GAG would continue to try to convince the companies of the practical value of deconvolution. From 1959 to 1961 I was a consultant to Pure Oil Company working under Howard Slack. Later (1982) Slack wrote, "We had you at Pure Oil, and if the company wasn't disbanded, then we would have caused the digital revolution instead of Amoco." In 1962 I was a consultant to Amoco working with Sven Treitel. Amoco became the first oil company to have a fully operational deconvolution system. Shell was not far behind. In 1964 while I was with of Geoscience Inc. Shell gave Geoscience a large contract to expand their deconvolution system. By 1965, the entire oil industry was converting from analog to digital seismic processing. In 1965 David Brown, Rudy Prince, David Steetle, George Cloudy, William Shell, Pat Poe, and I founded Digicon, an independent exploration company. In 1968 three of the four Digicon land seismic crews were operating on the Alaskan North Slope, while two of Digicon's four marine vessels (Figure 11) were in the Beaufort Sea (the portion of the Arctic Ocean north of Alaska and Yukon).

Figure 11. In 1969, the *Gulf Seal*, a 165 foot vessel leased by Digicon, equipped for making seismic surveys in the Gulf of Mexico.

In the mid-1960s, 83 percent of Raytheon's business was with the government. During the late 1960s and 1970s, however, Raytheon began diversifying from defense into other fields such as publishing, home appliances and energy. In 1966 Raytheon bought Seismograph Service, an international geophysical company. Seismograph Service prospered and Raytheon took its rightful place as a leading force in digital oil exploration.

The most decisive factor was a new generation of students knowledgeable in computers.   By the 1960s, MIT and other universities were turning out students who knew digital computing from the ground up. None of these students had to be convinced that the digital computer was here to stay. They were enthusiastic. The oil companies and geophysical companies were quick to hire recent graduates. They newly hired scientists read the MIT Geophysical Analysis Group Reports and took up where we left off in 1954. Things happened quickly. By the 1970s the entire oil exploration industry was almost exclusively digital.

"If you can look into the seeds of time, and say which grain will grow and which will not, speak then to me" (Shakespeare). Great strides were continually being made in digital computers. With the introduction of transistorized computers, like the IBM 7090 and the Control Data 1604, both computer technology and computer applications were in the ascendancy. In 1959 Jack Kilby of Texas Instruments patented the first integrated circuit (IC). Robert Noyce at Fairchild sold the first commercial IC chips in quantity in 1961. The conversion to digital seismic processing did start to happen in the early 1960s and exploration geophysics became the first of any science to experience the digital revolution. The oil and geophysical companies began using deconvolution and other digital signal processing methods on all exploration seismic records. Provinces for oil exploration that yielded only no-reflection seismograms were now unlocked. Included were great offshore prospects whose seismograms were typically clouded by the reverberations of the water layer. Raytheon, through Seismograph Service Company and the work of Dale Stone, Bob Geyer and others, took its rightful pace as a leader in seismic processing. The enthusiasm for digital seismic processing has continued uninterrupted to this day. The changes wrought in geophysical exploration are comparable to the changes wrought in astronomy with the invention of the telescope. The exciting story of exploration geophysics from its early beginnings is well told in the excellent and comprehensive book of Lawyer, Bates, and Rice (2001). The success of reflection seismology in discovering petroleum depends upon accurate images of the interior of the earth. Digital processing makes these images possible.

Since 1964, deconvolution together the associated digital imaging has been responsible for the discovery of a trillion barrels of new oil as well as great amounts of natural gas. With oil and gas taking the place of coal, lives lost from cave-ins in underground coal mines have been reduced. Northern cities are no longer subjected to black soot from the burning of coal. Natural gas from the oil fields in the North Sea did away with the London fog as described in the Sherlock Holmes stories. Oil exploration is no longer the only enterprise completely dependent upon the digital computer. Today the computer is everywhere: manufacturing, transportation, medical care, food production, communications. Shakespeare wrote, "Banish plump Jack, and banish all the world." Today it would be: "Banish the computer, and banish all the world."

Bill Mueller

> Thursday, July 10, 2008
> Enders,
> I owe you a debt of gratitude I don't believe I ever re-paid. In the middle 60's I was a geophysics major at the University of Tulsa under Parke Dickey and Eysteinn Tryggvason. Most of my geophysics courses were taught at night by research geophysicists from Pan American Research center in Tulsa. [Added note: Pan American (Amoco) was later acquired by BP]. I understood that they all worked with you and they brought us the cutting-edge work that you were doing at the time on signal processing, especially deconvolution.

I was way, way ahead of most of the industry when I graduated
and it was that background helped me immensely in starting a
geophysics career that is now approaching 40 years.

Again, Thanks
Bill Mueller, Antlers Exploration, "We See Things Others Don't"

Interactive earth-digital processing

In order to increase seismic resolution, we may call upon the methods
of interactive design. This discipline draws upon elements of the
interaction of two systems as a basis of increasing resolution. The most
celebrated form of interactive design is the relationship of the artist
with his object. No clearer statement can be found than that given by
Michelangelo with the words

> With chiseled touch, the stone unhewn and cold becomes a
> living mold.
> The more the marble wastes, the more the statue grows.

In exploration the artist is the geophysicist, the computer is the chisel,
and the object is made up of the underground stone layers of the earth.
These layers can serve as a mold. The geophysicist must use the
computer to chisel away the layers, as it were, to reveal the statue,
namely a computer image of the subterranean earth.

Although the computer is at the center of interactive design, the
discipline of interactive geophysical processing is not the same as that
of conventional geophysical processing. Current geophysical processing
uses the computer to refine and image a fixed set of seismic data that
has been previously recorded in the field. Let us give an analogy from
photography. An artist would use an analog camera to take a picture of
a person. The picture was the final image. In the analog days of seismic
prospecting, a geophysicist would use a seismic crew to record a seismic
section. The seismic section was the final image.

In these digital days an artist uses a digital camera to take a picture of
the person. The computer was then used to enhance the digital picture

to obtain the final image. In present-day seismic prospecting, the data is first recorded in the field, and then later the computer is used to enhance the field seismic data to obtain the final image. In either case, a fixed set of data in used to obtain the final image. See Figure 12.

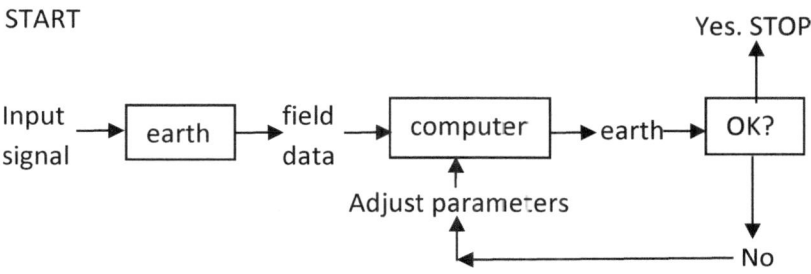

Figure 12. Present-day seismic processing

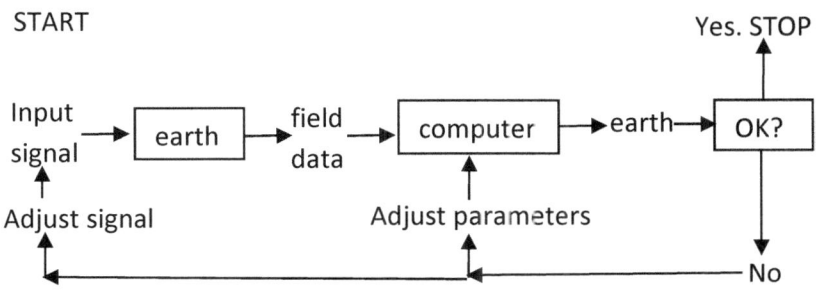

Figure 13. Interactive seismic processing

In Interactive processing, an artist uses a digital camera to take a picture of a person. The artist looks at the picture just taken. As feedback, the artist adjusts the camera and adjusts the pose of the person as well. A second picture is taken. The artist repeats this process until a satisfactory picture is produced. In interactive seismic processing, a input signal is coded both in time and in space. This coded signal is used to record a set of seismic field data. The field data is run directly into a computer on location. The computer analyzes the set of field data. As feedback, the computer adjusts the computer parameters and adjusts the coded signal as well. These feedback instructions occur in real time. A second set of seismic field data is taken. The geophysicist repeats this process until a satisfactory image of the earth is produced. See Figure 13. The essence of interactive design both in photography and in geophysics is the use successive sets of data produced by the object itself (i.e., the person or the earth, as the case may be) to refine the image in order to obtain better resolution.

The computer can be incorporated in the total environment of science. The earth itself becomes a computing machine which aids the deconvolution process as the signals are recorded. Thus, we can use the earth itself to analyze the data from the earth. In this way, we can utilize seismic recording systems more efficiently; our equipment will not be idle a good part of the time as it is now, but will be used at great efficiency by incorporating the environment itself into our computing system. In the biomedical sciences, the human body itself will be used as a computer under control of digital computers, and thereby the final data recorded will already be deconvolved and hence in a form ready for interpretation. Thus, we make the environment act as a computer which in turn is linked to the digital computer. Energy is recycled into the environmental computer in accordance with the needs determined by using signals returned from the environment. It is possible for us to see the results of our own ideas at the time of recording by such use of the environment as a computer. The information from the environment can be unfolded or deconvolved during the recording process itself, for we can link the digital computer to the environmental computer to acquire such information.

Seismic exploration is made up of two phases, which may be called the deconvolution (or analysis) phase and the re-convolution (or synthesis) phase. In the deconvolution phase the field data (a convolutional model) is pulled apart so the various components may be separated. In particular the primary reflections are isolated from all the other components. In the re-convolution phase the primary reflections are pulled together by use of the Huygens-Fresnel construction to yield a three dimensional image of the subsurface rock layers. In interactive seismic design, perceived flaws in the image are fed back to adjust and correct the input signatures as well as the parameters used in the computer processing. In other words, the iterative improvement process does not just work on one given set of seismic field data. Each iteration generates an entirely new set of fie d data. The earth itself enters in to the design, as it should.

The marriage of the earth with the computer is illustrated by directional drilling. In traditional drilling, an oilrig rotated the entire drill string. That method fails when a well is required to bend at an angle so sharp that the rotating string would break. Wells that bend a full 90 degrees over a space of a few hundred meters were made possible by the development of steerable drill strings. Such strings do not rotate but instead a mud-driven motor located at the end of the drill string drives the drill bit. Sensors near the bit are used to determine the porosity and density of the nearby rock. Thus it may be said that the earth itself determines the direction of the drill. In an analogous way, for each iteration of an interactive design, the earth itself determines the seismic data collected. Only by getting close to the earth can seismic resolution be significantly improved.

Lee Lawyer

After a degree in geological engineering at the University of Oklahoma and three years in the U.S. Army Corps of Engineers, Lee Lawyer joined Standard Oil Company of Texas (an ancestor of Chevron Corporation) in 1954. He would remain with Chevron until retirement 38 years later. That time frame meant a front-row seat in the transformation of seismic technology from analog paper records interpreted primarily in the field (often on the hood of the seismic party chief's pickup) to computer-massaged 3D digital data interpreted on ultra-sophisticated workstations.

Lawyer was rewarded with the discovery of major oilfields, one of which (Mills Ranch) held the record for deepest productive field (25,000 feet) for several years. Lawyer was "intimately involved at all levels, from field practitioner through various supervisory positions to executive-suite rococo as Chevron's chief geophysicist." Lee Lawyer's volunteer work with the Society of Exploration Geophysicists (SEG) has been virtually nonstop and continues today. He served as President of the SEG in 1987-1988. Shortly after retiring from Chevron, Lee has written a column for the geophysical journal *The Leading Edge* (TLE). His column appears in every issue. It is universally popular and remarkably prescient. The following is an email.

> Date: Sun, 14 Aug 2005 From: L. LAWYER To: Enders Robinson Subject: GAG
>
> Enders, I have just finished reading your article on GAG in the journal *Geophysics*. What a great story. I truly think it should be mandatory reading for all geophysicists. Well, maybe not mandatory, but high on the list of suggested reading. I liked the statement that the convolutional model turned the seismic world upside down. Fortunately, I was not in the seismic world at that time [1950-1954] so I didn't suffer any *mal de mer* as a result, but I get the point. Addition becomes convolution, signal becomes noise, noise becomes signal and analog processing becomes digital processing. Wow!
>
> Your article brought it into focus the analog processing "era" that we went through. I had almost forgotten the frequency slicing we did. I have forgotten whether the results were worth all of that effort and film! Aside: Do you remember all of that silver we recovered when developing film? I think we sold it and added the money to the employee annual party, or something like that. Lee

Professor N. L. Mohan

May 6, 1998
Dr. N.L. MOHAN, Ph.D.
PROFESSOR OF GEOPHYSICS, CENTRE OF EXPLORATION GEOPHYSICS
OSMANIA UNIVERSITY, HYDERABAD-500007, INDIA

To Prof. Enders A. Robinson, Professor of Geophysics,
Columbia University, New York, USA

Dear Prof. Robinson,

Please find the reprint, *Fractality of Seismic Wave Signatures*. "Why did this an unknown man dedicate this article to me?" would be indeed a natural question that might arise in your mind. A thought always has been lingering in my mind that I should have become Ph.D. student of Prof. Robinson. ...

In this context here is an interesting story from our Indian Epic "MAHABHARATA", an paralleled classic, believed to be more than 5000 years old, that envelopes the manifestation of human mind that runs simultaneously both on down to earth material plane and highest philosophical plane. Here is an interesting story from the Epic "Mahabharata":

The king of Hastinapur (Now it is New Delhi, the capital of India) appoints a teacher, Dronacharya, for imparting the knowledge to his sons in every field, and particularly to make them powerful warriors. He takes them to a forest and starts imparting knowledge to the king's sons. One son called Arjuna impresses the teacher, Dronacharya, well because of his extraordinary concentration, skill to hit the objects using the bow and the arrow with right and left hands that too in darkness by sensing the sound, above all his obedience. He promises Arjuna that he will make him a great warrior and see that no one could conquer him on this earth.

One day a tribal boy comes to Dronacharya (while he is giving training to the king's sons) and requests him to take him as his student as he also wants to learn like those sons. But he refuses to impart him knowledge. Without losing heart, the tribal boy makes a statue of Dronacharya with sand and mud and keeps under the banyan tree and worships him as his teacher and starts learning on his own. In course of time he becomes a great warrior, more than the king's son, Arjuna. After some months, the teacher and the taught move around the forest and encounters this tribal boy and his (teacher's) image beneath the tree, hitting the object using the bow and the arrow with great precision. The tribal boy immediately welcomes Dronacharya with great respect and reverence, offers him fruits etc. and tells him how he started learning different skills. Dronacharya wonders at his (tribal boy) power of concentration, intelligence, extraordinary skills and knowledge, and also make Arjuna to shiver.

It is an age old Indian tradition that after the completion of the course the student offers the teacher some presentation called "GURU DAKSHINA" (Guru = teacher; Dakshina = a present to the teacher for imparting knowledge, now we call fees). Of course, it depends on one's own capacity to offer and never the teacher demands. Following the tradition, the tribal boy offers Dronacharya a gift according to his wish (as he worships him as his teacher though he never taught him). At that moment the King's son, Arjuna, reminds Dronacharya about his promise. Then the teacher immediately asks him to offer his right hand thumb finger as "Guru Dakshina" and without a second thought the tribal boy, with great humility, cuts his right hand thumb finger, knowing fully well that he will be unfit to continue as warrior, and places at his feet. Then onwards he becomes famous as "EKALAVYA" (a man who learns on his own).

I am quite sure you would appreciate the deeply embedded philosophy of the above said story, more than the material aspect. ... I thoroughly enjoyed your two articles appeared in TLE (1997, 1998), one pertaining to the elimination of source signature using the z-transformation and thereby obtaining the reflection coefficients; and the other one on holistic migration. ...

With kind regards and thanks, Yours sincerely, N. L. MOHAN

Nigel Anstey

In 2014 the European Association of Geoscientists and Engineers (EAGE) awarded the great English geophysicist Nigel Anstey its highest honor. The citation is:

> The Erasmus Award is for a lifetime contribution of outstanding and lasting achievements; against such a yardstick, few can match the achievements made by Nigel Anstey over seven

decades. Nigel has made lasting contributions to seismic theory, acquisition, data processing and interpretation. As a one-man asset team, Nigel was an early destroyer of the artificial boundaries between geology and geophysics to the benefit of generations of young earth scientists who have read, enjoyed and learned from his books and teaching materials. Nigel has made outstanding contributions to resource exploration and development, ranging from his seminal papers on seismic wavelets; the statistics of thin beds and their effect on the propagating wavelet; the invention with Bill Lerwill of the magnetic correlator which made the emerging Vibroseis method a practical technique; a patent on vertical seismic profiling, and the introduction of coloured overlays of seismic attributes in interpretation.

There is practically no area of seismic acquisition, processing or interpretation that has not been touched by Nigel's contributions. Not only has he been a great innovator but Nigel has a gift for writing that demystifies concepts and explains them, to novices and experienced practitioners alike, to reveal fresh insights and understanding. This gift has spurred Nigel to write excellent books on Vibroseis, seismic acquisition and seismic interpretation, as well as materials for the many courses and lectures that he has delivered over the years. Most of today's luminaries studied with Nigel's books and course notes, or had the privilege to hear his lectures in person, and it is an honour for EAGE to present its most prestigious award to such a worthy recipient.

Nigel Anstey wrote the following citation in *The Leading Edge*, Volume 3, Issue 4, April 1984, Society of Exploration Geophysicists, Tulsa, OK:

If each of us was asked to nominate the Great Leaps Forward in our science, and to associate each such breakthrough with a name, all our lists would have one common entry: deconvolution, and the name of Enders A Robinson. Today

[1983], in recognition of his immense service to geophysics, the Society of Exploration Geophysicists (SEG) adds that name to the rolls of the Honorary Members of the Society.

In September of 1950, as a young graduate student at MIT, Enders Robinson commenced the task which would revolutionize seismics. It must have seemed very humdrum at the time; it was the digitization, with a ruler and pencil, of eight seismic records from Texas. By the spring of 1951 there were autocorrelations and spectra. No surprises there—it was deconvolution which was the great unknown...Would it work on real data? It took the whole summer of 1951 to deconvolve 32 traces. The first trace (plotted, of course, by hand) looked too good to be true. But then the second, and the third... This success led to the establishment of the Geophysical Analysis Group at MIT; Enders Robinson was its first Director. The work of this group (1952-1957) had three profound effects on geophysics:

— First and foremost, of course, it developed the technology of deconvolution. Although scores of researchers have since written hundreds of papers on the subject of deconvolution, almost all of the basic techniques used today remain those set out by Enders Robinson in his classic papers of the 1950's.

— It forced the digital revolution. For corrections and stacking and filtering—perhaps even for velocity analysis and dereverberation—the industry could have continued to make progress by analog means; for statistical deconvolution the digital route was virtually the only choice.

— It created a totally new respect for theory. Seismic prospecting, to that time, had been very much a practical endeavor; doodlebuggers had scant regard for mathematicians. But deconvolution worked, and deconvolution came from theory; now there was no doubt that we must listen when theory spoke.

How fortunate we were, then, that theory spoke in the person of Enders Robinson! For Enders could communicate. True, many of us would grimace a little when Enders would say in the preamble that we would need only elementary algebra, for we knew that what was elementary algebra to Enders was the threshold of pain for the rest of us. But it was always worth the effort; each knotty development would be followed by a clear verbal summary, and at the end we would understand. Thus was a whole generation of geophysicists reoriented by the writings of Enders Robinson (joined, from time to time, by a happy choice of coauthors).

Enders output has been phenomenal. In more than 60 papers and essays, and in more than 20 books, he has guided and chronicled the evolution of signal processing from the hand digitization of the 1950's to the custom deconvolution chip of the 1980's, while also stimulating the adoption of these techniques in radar, speech analysis, economics and many other sciences. Thus has Enders provided for us the all-important bridge between mathematics and applied physics— without which the theory is an abstraction and the practice is unfulfilled. And to all of this Enders brought an eminently readable style and an infectious delight in the beauty of science.

Of course, Enders would be the first to remind us that he stands on the shoulders of the giants of the past. Nevertheless, to a whole generation of geophysicists—and almost despite his major contributions to other aspects of signal processing—he will always remain the Father of Deconvolution.

Epilog

A few days before Christmas in 2003, my wife, her brother, and I met MIT Professor Jay Forrester and his wife at breakfast at the Clark Currier Inn in Newburyport, Massachusetts. Jay Forrester was the person responsible for the invention and the development of the MIT Whirlwind computer, begun in 1945 and first demonstrated on April 20,

1951. We talked about old times. We discussed the delay-line memory (RAM) used in the Eckert Mauchly computers. These computers were simple but slow. Forrester mentioned Howard Aiken, who built a series of computers at Harvard starting in 1944. Aiken had a reputation of being stubborn. Forester said that he may have been the only one who was not thrown out of Aiken's office. We talked about the electrostatic memory that was used in the first configuration of Whirlwind. The entire electrostatic memory had to be replaced every month, and it cost $1 per bit. The lack of any kind of dependable random access memory (RAM) was the Achilles' heel of all existing computers. Forrester's invention of solid-state core memory was the breakthrough that made the digital computer a practical device. Forrester was alone in the development of core memory; nobody wanted any part of it. After it was invented, everyone took credit for it. Forrester lamented, "Success has a thousand fathers; failure in an orphan." The conversation went on. As we were leaving, Forester told me, "You are one who understood." Forrester always considered digital geophysics as the prime example of a field that got started on Whirlwind. At that moment, the tug of past days at MIT was strong. Charles Lamb wrote, "I was bound to traverse, seeking to find the old familiar faces. All are departed. All, all are gone, the old familiar faces."

Whirlwind 1

A HIGH-SPEED ELECTRONIC DIGITAL COMPUTER

www.ingramcontent.com/pod-product-compliance
Lightning Source LLC
Chambersburg PA
CBHW070232190526
45169CB00001B/161